T0317700

Principles of Modeling and Simulation

Principles of Modeling and Simulation

A Multidisciplinary Approach

Edited by

John A. Sokolowski
The Virginia Modeling, Analysis and Simulation Center
Old Dominion University
Norfolk, VA

Catherine M. Banks
The Virginia Modeling, Analysis and Simulation Center
Old Dominion University
Norfolk, VA

WILEY

A JOHN WILEY & SONS, INC., PUBLICATION

Published by John Wiley & Sons, Inc., Hoboken, New Jersey.
Published simultaneously in Canada.

For general information on our other products and services or for technical support, please contact our Customer Care Department within the United States at (800) 762-2974, outside the United States at (317) 572-3993 or fax (317) 572-4002.

Wiley publishes in a variety of print and electronic formats and by print-on-demand. Some material included with standard print versions of this book may not be included in e-books or in print-on-demand. If this book refers to media such as a CD or DVD that is not included in the version you purchased, you may download this material at http://booksupport.wiley.com. For more information about Wiley products, visit www.wiley.com

Library of Congress Cataloging-in-Publication Data:

Sokolowski, John A., 1953–
    Principles of modeling and simulation : a multidisciplinary approach / John A. Sokolowski, Catherine M. Banks.
        p. cm.
    Includes bibliographical references and index.
    ISBN 978-0-470-28943-3 (cloth)
  1. Mathematical models.   2. Simulation methods.   3. Interdisciplinary research.   I. Banks, Catherine M., 1960–   II. Title.
    QA401.S648 2009
    003—dc22
                                                                    2008021381

This book is dedicated to

Marsha, Amy, and Whitney

—John A. Sokolowski

Benjamin M. Mercury

—Catherine M. Banks

Contents

Preface

The impetus for this study is the realization that no textbook exists that provides an introduction to modeling and simulation suitable for multiple disciplines, especially those that are outside the science and engineering fields.

Many universities are realizing that modeling and simulation is becoming an important tool in solving and understanding numerous and diverse problems. They have begun to offer introductory courses in this field to acquaint their students with the foundational concepts that will help them apply modeling and simulation in many areas of research. This text serves to provide an orientation to the theory and applications of modeling and simulation from a multidisciplinary perspective.

To students who will be reading this text we offer a concise look at the key concepts making up the field of modeling and simulation. While modeling and simulation necessarily entails mathematical representations and computer programs, students need only be familiar with math at the college algebra level and the use of spreadsheets to understand the modeling and simulation concepts covered in this book.

The text is divided into three parts with nine chapters. Part One, *Principles of Modeling and Simulation: A Multidisciplinary Approach* introduces modeling and simulation and its role. Chapter 1 answers the question, "What Is Modeling and Simulation?" The chapter provides a brief history of modeling and simulation, lists the many uses or applications of modeling and simulation, and speaks to the advantages and disadvantages of using models in problem solving. Chapter 2 focuses on "The Role of Modeling and Simulation." It covers the two main reasons to employ modeling and simulation: solving a specific problem, and using modeling and simulation to gain insight into complex concepts.

Part Two, *Theoretical Underpinnings*, examines the most fundamental aspects of modeling and simulation. Chapter 3, "Simulation: Models That Vary over Time," provides a definition for simulation and introduces the reader to two main simulation concepts: discrete event simulation and simulation of continuous systems. Chapter 4, "Queue Modeling and Simulation" examines queuing models, sequential simulation, and parallel simulation. Chapter 5, "Human Interaction with Simulations," explains the two primary methods in which humans interface with simulations: simulation and data dependency and visual representation. Chapter 6, "Verification and Validation," answers two fundamental questions: What is verification and validation? and Why is verification and validation important?

Part Three, *Practical Domains*, affords the student an opportunity to consider the many uses of modeling and simulation as a tool in workforce development. Students will also review case studies and research conducted in various disciplines emphasizing the notion that models serve as approximations of real-world events.

Chapter 7 addresses the "Uses of Simulation." These uses are found in training, analysis, decision support, and acquisition. Chapter 8, "Modeling and Simulation: Real-World Examples," delves into specific applications of modeling and simulation in the Transportation, Business, Medical, and Social Sciences domains. Chapter 9 addresses "The Future of Simulation" by providing a basis for accepting modeling and simulation as a discipline with its own body of knowledge, research methods, and curriculum of study. It provides answers to the questions: Is modeling and simulation a tool or discipline? How should education, research, and training be conducted to support workforce development?

Contributors

Michael P. Bailey
United States Marine Corps
Operations Analysis Division
Quantico, Virginia

Catherine M. Banks
Virginia Modeling, Analysis and
    Simulation Center
Old Dominion University
Norfolk, Virginia

C. Donald Combs
Planning and Health Professions
Eastern Virginia Medical School
Norfolk, Virginia

David P. Cook
Department of Information Systems
    and Decision Sciences
Old Dominion University
Norfolk, Virginia

Paul A. Fishwick
Department of Computer and
    Information Science and
    Engineering
University of Florida
Gainesville, Florida
    and, **Hyungwook Park**
    Graduate Research Assistant

Michael D. Fontaine
Virginia Transportation Research
    Center at
University of Virginia
Charlottesville, Virginia

R. Bowen Loftin
Vice President and CEO
Texas A&M University
Galveston, Texas

Ahmed K. Noor
Center for Advanced Engineering
    Environments
Old Dominion University
Norfolk, Virginia

Tuncer I. Ören
Information Technology Research
    Institute (ITRI) / Engineering
University of Ottawa
Ottawa, Canada

Mikel D. Petty
Center for Modeling, Simulation, and
    Analysis
University of Alabama—Huntsville
Huntsville, Alabama

Paul F. Reynolds, Jr.
Department of Computer Science
University of Virginia
Charlottesville, Virginia

John A. Sokolowski
Virginia Modeling, Analysis and
    Simulation Center
Old Dominion University
Norfolk, Virginia

Part One

Principles of Modeling and Simulation: A Multidisciplinary Approach

Chapter 1

What Is Modeling and Simulation?

Catherine M. Banks

INTRODUCTION

Modeling and Simulation, or M&S as it is commonly referred, is becoming one of the academic programs of choice for students in all disciplines. M&S is a discipline with its own body of knowledge, theory, and research methodology. At the core of the discipline is the fundamental notion that *models are approximations for the real-world*. To engage M&S, students must first create a model approximating an event. The model is then followed by *simulation*, which allows for the repeated observation of the model. After one or many simulations of the model, a third step takes place and that is *analysis*. Analysis aids in the ability to draw conclusions, verify and validate the research, and make recommendations based on various iterations or simulations of the model. These basic precepts coupled with *visualization*, the ability to represent data as a way to interface with the model, make M&S a problem-based discipline that allows for repeated testing of a hypothesis. Teaching these precepts and providing research and development opportunities are core to M&S education. M&S also serves as a tool or application that expands the ability to analyze and communicate new research or findings.

There has been much attention paid to M&S by the National Science Foundation (NSF). In 1999, then Director Dr. Rita R. Colwell declared simulation as the *third branch* of science at the fall meeting of the American Geophysical Union [1]. In a more recent report entitled, "Simulation-based Engineering Science: Revolutionizing Engineering Science through Simulation," the NSF drew on the expertise of an esteemed cadre of scientists to discuss the challenges facing the United States as a technological world leader. The group made four recommendations that they believed would help restore the United States to its leadership role in this strategically critical technology (simulation). One recommendation went straight to the study of M&S:

> *"The Panel recommends that NSF underwrite an effort to explore the possibility of initiating a sweeping overhaul of our engineering educational system to reflect the multidisciplinary nature of modern engineering and to help students acquire the necessary modeling and simulation skills."* [2]
>
> Simulation-Based Engineering Science: Final Report, May 2006

The intent of this text is to introduce you to M&S education and research from a multidisciplinary approach so that you can acquire the skills necessary to this critical technology.

Fundamental to a formal engineering M&S program of study is its curriculum built upon four precepts—modeling, simulation, visualization, and analysis. Students study the basics of **modeling** as a way to understand the various modeling paradigms appropriate for conducting digital computer simulations. They must understand **simulation** and the methodology, development, verification and validation, and design of simulation experiments. Students who are able to engage **visualization** are able to provide an overview of interactive, real-time 3D computer graphics and visual simulations using high-level development tools. Important to any student research is the **analysis** of the findings; and included in any good analysis is an observation of the constraints and requirements of applying M&S. In other words, analysis also includes making known the limitations of the research.

It was political scientist Herbert A. Simon (1916–2001) who introduced the notion of *learning by doing* (also known as experiential learning).[1] M&S can be just that. It is the simulation of a model that allows for the imitation of the operation of a real-world process or system over time. To imitate an operation over time one must generate a history, real or artificial, to draw inferences concerning the operating characteristics of the real system that is represented [3]. The art and science of M&S has evolved very rapidly since the mid-1980s, so much so that it easily parallels the technological advances of mainframe and desktop computers and the ever-increasing emergence of the internet and World Wide Web (www).

---

[1] **Herbert A. Simon** was a political scientist who conducted research in a variety of disciplines including cognitive psychology, computer science, public administration, economics, management, and philosophy of science. Dr. Simon was among the founding fathers of several of today's most important scientific domains, including artificial intelligence, information processing, decision-making, problem-solving, attention economics, organization theory, complex systems, and computer simulation of scientific discovery. He was the first to analyze the architecture of complexity and to propose a preferential attachment mechanism to explain power law distributions. He introduced the notion of *experiential learning, bounded rationality*, and *satisficing*. Dr. Simon's research at Carnegie Mellon University resulted in numerous cited publications. He remains one of the most influential social scientists of the 20th century.

MODELS: APPROXIMATIONS OF REAL-WORLD EVENTS

A *model* is a representation of an event and/or things that is real (a case study) or contrived (a use-case). It can be a representation of an actual system. It can be something used in lieu of the real thing to better understand a certain aspect about that thing. To produce a model you must abstract from reality a description of a vibrant system. The model can depict the system at some point of abstraction or at multiple levels of the abstraction with the goal of representing the system in a mathematically reliable fashion. A *simulation* is an applied methodology that can describe the behavior of that system using either a mathematical model or a symbolic model [4]. Simply, simulation is the imitation of the operation of a real-world process or system over a period of time [3]. As you will see there are many uses of M&S. M&S can be used to determine the ordering policies of Wal-Mart's extensive inventory system, or it can be used to analyze the prospects and rate of rehabilitation of a patient who just underwent knee-replacement surgery, or it can be used to evaluate ocean currents and waves to better understand weather patterns.

M&S begins with 1) developing computer simulation or a design based on a model of an actual or theoretical physical system, then 2) executing that model on a digital computer, and 3) analyzing the output. Models and the ability to act out with those models is a credible way of understanding the complexity and particulars of a real entity [4]. From these three steps you can see that M&S facilitates the simulation of a system and then a testing of the hypothesis about that system. For example, if you wanted to determine how many cashiers are needed to process a certain number of customers during rush hour with the assurance that the store's high level of quality service was not compromised, you must first research the current system of processing customers.

You will no doubt review the work schedule and note that the manager has scheduled more cashiers during peak times. You will then assess how many customers are processed during peak times based on the cashier tapes. Also, you might want to see how long it takes to process a customer at slow periods and at heavy traffic periods—you might be surprised to find that customers are processed in shorter exchanges at busy times. Do the customers feel rushed? How many errors are made? Do the customer lines flow smoothly? Are the cash registers placed in good locations? All of this is part of the initial research you will do to develop your model. Once you have sufficient data you can create your model. It is important to note that *models are driven by data* and so your data collection must be done with great accuracy.

Once the model is created you can craft a fairly well-thought-out and credible hypothesis such as, *if the store manager does this, this will be his result.* But are you certain? There may be unexpected changes to the model—a cashier is out sick, a cash register breaks, the power goes out and stops all transactions. What can the manager do to accommodate these unforeseen occurrences? You can assist the manager by creating a number of simulations or iterations of the model to ascertain the "what if." Upon reviewing the output of your simulations, you can provide that data to the store manager so that he or she can make well-informed decisions about the scheduling of cashiers and distribution of registers to meet the goal of high-quality service. As you can see from the example, M&S gives you many opportunities to repeat a simulation of the hypothesis. In essence, you have the ability to repeat the testing of the hypothesis through various simulations. Let's take a closer look at simulation.

First, we must appreciate that defining simulation is not as clear-cut as defining model. Definitions of simulation range from:

- a method for implementing a model over time

- a technique for testing, analysis, or training in which real-world systems are used, or where real-world and conceptual systems are reproduced by a model

- an unobtrusive scientific method of inquiry involving experiments with a model rather than with the portion of reality that the model represents

- a methodology for extracting information from a model by observing the behavior of the model as it is executed

- a nontechnical term meaning not real, imitation (the correct word here is the adjective simulated)[2]

Simulation is used when the real system cannot be engaged. The real system may not be engaged because 1) it may not be accessible, 2) it may be dangerous to engage the system, 3) it may be unacceptable to engage the system, or 4) the system may simply not exist. So to counter these objections a computer will "imitate" operations of these various real-world facilities or processes. Modeling depends on computational science for the visualization and simulation of complex, large-scale phenomena. These models may be used to replicate complex systems that exhibit chaotic behavior and so simulation must be used to provide a more detailed view of the system. Simulation also allows for virtual reality research whereby the analyst is immersed within the simulated world through the use of devices such as head-mounted display, data gloves, freedom sensors, and forced-feedback elements [4]. *Artificial Life* and *Computer Animation* are offshoots of computational science that allow for additional variations in modeling.[3]

Now that we know what comprises a model and what constitutes a simulation, we then couple these steps with visualization. M&S coupled with visualization refers to the process of developing a model of a system, extracting information from the

---

[2] Additional information and definitions can be found at the U.S. Department of Defense, Defense Modeling and Simulation Office (DMSO) online glossary at http://www.dtic.mil/whs/directives/corres/pdf/500059m.pdf.

[3] *Artificial life* enables the analysts to challenge the experiment by allowing the computer program to simulate artificial life forms. *Computer animation* is emphasized within computer graphics and it allows the modeler to create a more cohesive model by basing the animation on more complex model types. With the increased use of system modeling there has been an increased use of computer animation, also called physically based modeling [4].

model (simulation), and using visualization to enhance our ability to understand or interpret that information. We have mentioned "system" a number of times. Let's take a look at what constitutes a system.

An accepted definition of "system" was developed by the International Council of Systems Engineering (INCOSE). A **system** is a *construct or collection of different elements that together produce results not obtainable by the elements alone.*[4] The elements can include people, hardware, software, facilities, policies, documents—all things required to produce system-level qualities, properties, characteristics, functions, behavior, and performance. Importantly, the value of the system as a whole is the relationship among the parts. There are two types of systems: 1) *discrete* in which the variables change instantaneously at separate points in time and, 2) *continuous* where the state variables change continuously with respect to time. There are a number of ways to study a system:

- the actual system versus a model of the system
- a physical versus mathematical representation
- analytical solution versus simulation solution (which exercises the simulation for inputs in question to see how they affect the output measures of performance) [5]

As you will learn, M&S holds a significant place in research and development due to its inherent properties of modeling, simulating, analyzing, and visualizing (communicating). It is becoming the training apparatus of choice. In fact, M&S is considered a new tool of choice in the fields of health services, education, social sciences, business and industry. Many folks in the M&S community (researchers, academicians, industry, and military) were introduced to M&S as a tool that evolved with the modern military of the 20th century. But its origins can be traced to an ancient military whose use of wargames made it one of the most efficient armies in military history.

A BRIEF HISTORY OF MODELING AND SIMULATION

The act of wargames and challenging or outwitting an opponent on the battlefield is centuries old. In ancient Rome, the then world's largest empire was secured by the world's largest military. The Roman Army conducted live training between two contingents of its military (red team versus blue team). Their training battlefield reflected an environment the troops would encounter somewhere within the expansive Roman Empire that spanned the Scottish border in northern Europe throughout North Africa into the Near East (Afghanistan). The Roman Army had to learn how to fight in unknown regions against armies with diverse warring techniques. Although their training exercises were not intended to draw blood, their training honed a mili-

---

[4] Additional information and definitions can be found at the INCOSE online glossary at http://www.incose.org/mediarelations/glossaryofseterms.aspx.

tary prowess that made the Roman Army the greatest military the world had known for centuries (circa 500 B.C.E.–1500 C.E.). Significant as they were with Rome's legions, models were not restricted to the art of wargames and military training.

During the age of the Renaissance (1200–1600 C.E), or rebirth of learning, artists and scientists were using models in their designs of statuary or edifices. These models were presented to the artist's patron or commissioner as a way of seeking approval of a design before beginning an expensive project such as a marble bust or sarcophagus. One of the most notable scientists of the time was Leonardo DaVinci. He is famous for his paintings, sculptures, building designs, and scientific experiments. His projects include the design of advanced weaponry (to include tanks and submarines), flying machines, municipal construction, canals, and ornamental architecture (churches and fortresses), as well as his famous anatomical studies. Among his many assignments Leonardo worked as a military engineer where he was called upon to design a bridge to span the Golden Horn (a freshwater waterway dividing the city) in Constantinople (modern day Istanbul). Leonardo was also commissioned to do a life-sized equestrian statue (which was later changed to be four times larger). To do this he studied the movement of horses, made countless sketches, and devised new casting techniques. He did not complete the project, but he had succeeded in making a 22-foot clay model.

Leonardo made repeated uses of modeling to test the design of many of his inventions and projects. He determined that by understanding how each separate machine part functioned, he could modify it and combine it with other parts in different ways to improve existing machines or create new machines. He provided one of the first systematic explanations of how machines work and how the elements of machines can be combined. Fortunately, his studies and sketches have been collected into various codices and manuscripts that are available for our review. Around this same time a new competition was introduced in Europe. It came in the form of a game that required intellect and prowess—chess.

The current game of chess, as most westerners know it, had its origins in southern Europe in the second half of the 15th century. That game was a derivation of a 7th century game of Indian origin. Included on the chessboard are the King, Queen, Bishop, Knight, Rook, and Pawn. The object of the game is to checkmate the opponent's King by first placing that King under immediate duress or "check" with such maneuvering that there is no way for your opponent to remove his King from attack. Think about what is created on the chessboard: a simulated battlefield with two armies who possess equal strength of force. It is now up to the human commander (the chess player) to conduct his simulations: what if I move this way? What will happen if I do this? How will my opponent respond? What is he planning? The ultimate "checkmate" is rewarded by winning the war (the game). But what if you are playing a computer? Can it outwit a human opponent? Yes, it can.

In 1997 an IBM chess playing computer named Deep Blue won a short six-game exhibition match (not a world title match) by two wins to one with three draws against the Russian world champion Garry Kasparov after he made a blunder in the opening of the last game. Kasparov accused Deep Blue—IBM—of cheating and demanded a rematch, but officials at IBM declined. His accusation stemmed from the fact that he saw deep intelligence and creativity in the machine's moves suggesting that during the second game human chess players, in violation of the rules, intervened. IBM's response was that the human intervention occurred only between games as the rules provided so the developers could modify the program between games. This gave IBM an opportunity to modify for weaknesses in the computer's play as it was displayed during the game. Doing these modifications precluded the computer from falling into previous traps set by Kasparov [6]. There are a number of theoreticians who have developed extensive chess strategy and tactics. Many who play the game cite chess as one of the first wargames. By the 18th century military modeling, simulation, and training took on a new perspective.

In the 1780s, with England at the height of its naval power, a Scotsman named John Clerk developed a method of using model ships to gain tactical insights. He used his ships to step through battles analyzing the influence that geometry of the combatants had on their combat power. While a military simulation, Mr. Clerk's work was not considered a wargame because it did not provide a way to measure or apply the effects of actions—the reward and risk from game theory [7]. On the European continent, however, wargames were being formally developed by the Prussians (modern-day northeastern Germany).

Prussia attained its greatest importance in the 18th and 19th centuries. In the 18th century during the reign of the Soldier King Frederick I (1713–1740), Prussia instituted a *standing army*, or an army composed of full-time professional soldiers who *stand over* and never disband, even during times of peace. As a result of this significant military capacity, Prussia became a great European power during the latter half of the century under the reign of Frederick II (1740–1786). The Prussians saw the advantages to playing wargames and by 1824 games were incorporated as part of the training throughout the Prussian army. It was during the 19th century that Prime Minister Otto von Bismarck pursued a policy of uniting the German principalities into a "Lesser Germany" that would exclude the Austrian Empire. This led to the unification of Germany in 1871. Wargaming no doubt contributed to the outstanding military capability of Prussia's standing army and its success on the battlefield during the 19th century.

In the United States, Major W. R. Livermore of the Army Corps of Engineers introduced modern wargaming to the American military [7]. In 1883 he translated the German rules to a wargame they had developed based on the American Civil War and Prussia's own wars of 1866 and 1870–1871. Livermore found that when he compared the German attrition tables to actual statistics errors were made. Livermore determined that the German attrition tables usually predicted lower casualties than the historical record indicated. He adjusted his tables accordingly. Upon improving the wargame with the historically accurate data, Livermore sought official acceptance of wargaming for the U.S. military. Much to his surprise he was blocked by General William T. Sherman who was serving as the U.S. Army's Chief-of-Staff. Sherman felt wargames depicted men as blocks of wood rather than human beings. He therefore refused the integration of wargames into military training. Four years after Sherman's refusal to use wargames, the Naval War College decided it would use Livermore's model. In 1887 the college introduced its first Army-Navy field exercise. By the turn of the 20th century wargames made their way into U.S. military exercises and training. These games, however, lacked the capability and the complexity to model an event with the accuracy we now see in military modeling, a change that came about with the introduction of technology.

M&S can trace its technical origins to the first flight simulator, the *Link Flight Simulator*, which was patented in 1929 by American Edward Link. The pilot trainer resembled a toy airplane from the outside with short wooden wings and fuselage mounted on a universal joint. Link used organ bellows driven by an electric pump to make the trainer pitch and roll as the pilot worked the controls. The *Link Flight Simulator* was used in both government and private sectors [8]. In 1931 *Link Flight Simulator* was fully instrumented and sold to the Navy. The Army took delivery of *Link* trainers three years later. And on the home-front, the *Link Flight Simulator* was used in amusement parks during the 1930s. This was no doubt great fun for those young at heart who enjoyed pretending to be a pilot. But importantly, the *Link Flight Simulator* was great economy for the military as vast sums of money and time were saved with the training of Navy and Army pilots in simulators replicating air flight. This is a good example of how using simulation allows the military or any other company or organization to test a system before investing in the full-scale model, or to train an individual in a less expensive environment.

In the post–World War I period the Navy and the Marine Corps both employed wargames as part of their training. This training proved useful with the coming of the Second World War. Under the leadership of General George C. Marshall live simulation was introduced into military training. As a result, military M&S was making quick inroads into training the military of a new world power and that was by no accident. With the end of two World Wars a new period of military engagement was beginning, one that brought with it weapons of mass destruction that required computer-assisted air defense systems to interfere with their delivery. This post–World War II period is called the *Cold War*. The Cold War took place between the two world powers: the United States and the Soviet Union, and it would last almost a half century (1945–1989) as a military competition between the two.

On 29 August 1949 the Soviet Union detonated an atomic device at the Semipalatinsk Test Site in Kazakhstan, making it the second nation in the world to detonate a nuclear mechanism. This action served as the impetus for the U.S. government to give grave consideration to the threat of another nation possessing nuclear military capability. As a result, the Department of Defense was given the approval to invest funds for research in air defense systems. By the winter of 1949 digital computers were engaged in creating simulated combat. Developed by the U.S. Air Force, a semi-automated ground environment—SAGE—simulated combat from the perspective of more than one combatant. This type of simulation provided military training that now incorporated an air-defense system.

By the 1950s computers were being used to generate model behavior followed by simulation programs. These computers were then required to process the results of the elements of the simulation-based problem solving environments [9]. Digital radar was now able to transmit from the newly developed Microwave Early Warning (MEW) radar. This innovative research was being conducted by engineers at the

Massachusetts Institute of Technology (MIT). Significant to the research was a transmission that tied together the MEW located at Hanscom Field to the digital computer named *Whirlwind* located at MIT. Also at this time a scaled-down version of SAGE was being developed. Dubbed the *Cape Cod System*, this simulator was introduced in 1953. It allowed radar operators and weapons controllers to react to simulated targets presented to them exactly as these targets would appear in an engagement. The country was becoming embroiled in a military contest that called for technology far beyond the imagination of the average citizen.

Interestingly, some of that same technology was making its way into the homes of so many families in everyday, ordinary appliances and communication devices that brought a new definition to the post-modern age. In essence, as the country was developing militarily, and so was every other aspect of technology—that is why the 1950s were so progressive. This was a unique time in the social history of the country. It was both an age of innocence and a post-modern world with technical advances that would send men into outer space. Ironically, it was the newly invented *RCA FlipTop* television and *Regency TR1* transistor radio that delivered fear and talk of war with the Soviets into the American family living room.

At the close of his two-term presidency (1953–1961), Dwight D. Eisenhower gave an address to the American people about the effects of the ongoing military competition with the Soviet Union. Eisenhower's Military-Industrial Complex Speech made Americans aware that *a vital element in keeping the peace is our military establishment*. The president emphasized that *U.S. arms must be mighty, ready for instant action, so that no potential aggressor may be tempted to risk his own destruction* [10]. To do this the federal government would support and fund research that would make the military state-of-the-art, always ahead of the opposition. The president's speech referred to the increasing military buildup in the United States throughout the 1950s. That build-up fueled the nation's growing economy and many were living quite comfortably during this time. Perhaps somewhat oblivious to what was truly happening, Eisenhower was compelled to explain to his fellow citizens the ramifications of coupling an immense military establishment with an expanding arms industry. This was a new concept for Americans. In fact, the **Military-Industrial Complex** was a new American experience with an economic and political influence that reverberated throughout the country. By 1960 the increased spending for this complex amounted to more than half of the U.S. federal expenditure. And, as the complex grew so did the workforce. From the close of World War II (1945) to the end of Eisenhower's second term (1961), an expansive workforce of civilian employees constituted much of the defense industry. Additionally, many universities thrived on the increased research opportunities.

Throughout the 1960s military wargames became much more sophisticated, moving from strictly tactical training to strategic commands. Games were now incorporating things like the political capacity of a state or leader. They also became technically mature. This became apparent with work done at the universities. In 1961 a student at MIT created an interactive computer game called *Spacewar* [11]. The game required the player to operate his spaceship during a conflict that was fought with the firing of torpedoes. Pilots of the spaceships scored points by launching missiles that inflicted damage on the opponent, avoiding direct hits by the opponent, and maneuvering the spaceship to avoid the gravitational pull of the sun. This computer game was one of the first *interactive* games in the country. The president and Congress were also pushing forward a research agenda at government institutions. Just over a decade after Spacewar, two engineers at the National Aeronautics and Space Administration (NASA) in Moffett Field, California, developed another computer game, one a bit more complex called *Mazewar* [12]. This game was networked and it introduced the concept of online players as **avatars** (a graphical image of a user or a graphical personification of a computer or a computer process) in the form of an eyeball chasing other players around a maze.[5] Mazewar's development in 1974 served as a catalyst for a number of versions on various programs.

The military was also making contributions to M&S by formalizing simulation as a training tool. In 1971 the Navy's *Top Gun* school opened to train fleet fighter pilots in air combat tactics. In 1975 the Tactical Advanced Combat Direction and Electronic Warfare (*TACDEW*) simulator was being used for team training. The simulator created 22 separate shipboard mock-ups with the ability to generate a virtual (also called synthetic) threat environment [13]. There was also work done with linking training simulators. Fighter plane cockpits like the B-52 (long-range, heavy bomber aircraft) were simulated so that they could operate with tanker (refueling aircraft) simulators to facilitate training plane/tanker rendezvous.

By 1983 simulator networking was advancing rapidly. The *Defense Advanced Research Projects Agency (DARPA)* had initiated simulator networking—*SIMNET*—with an emphasis on tactical team performance on the battlefield. The U.S. Army supported the idea of incorporating armor, mechanized infantry, helicopters, artillery, communications, and logistics into the model for a much more expansive simulated training experience. Combatants could now see each other and communicate over radios. The SIMNET simulator was introduced at the First-Platoon Level in 1986. By 1990 over 250 networked simulators at 11 different sites were delivered to the U.S. Army [14]. It wasn't long before the benefit of SIMNET training was realized.

On July 25, 1990, Saddam Hussein convened a meeting with U.S. Ambassador to Iraq April Glaspie expressing his contempt for two of his Persian Gulf neighbors, Kuwait and the United Arab Emirates. He specifically accused Kuwait of exceeding the Organization for Petroleum Exporting Countries (OPEC) production limits and thus driving down oil prices. This lowering of prices was having a negative affect on the Iraqi economy and he faulted the United States for encouraging this high level of production. Additionally, his aggressive behavior earlier in the year resulted in the cessation of American aid—no more American aid meant he would look elsewhere, and that elsewhere was Kuwait. Within two weeks of his meeting with Ambassador Glaspie, Saddam ordered his troops into Kuwait. Iraqi troops entered

---

[5] Please note the difference between a GAME and a SIMULATION. A game is more concerned with entertaining and there is much more player participation. A simulation is more focused on getting the details of the model and system correct. A simulation does not require a participant or player, but a game does.

Kuwait on August 2, six days later an international coalition that included U.S. ground forces was conducting *Operation Desert Shield* to counter the Iraqi invasion of Kuwait. By January 1991 the U.S.-led international coalition's mission changed to include offensive air attacks. A seamless transition from *Operation Desert Shield* to *Operation Desert Storm* was underway.

In February a decisive tank battle was in progress. The *Battle of 73 Easting* was fought between armored forces of the U.S. Army and the Iraqi Republican Guard. The U.S. ground unit was outnumbered and outgunned; yet, it was able to affect the enemy by destroying 85 tanks, 40 personnel carriers, and 30 wheeled vehicles carrying anti-aircraft artillery. Why was this outnumbered and outgunned U.S. armored unit able to conduct itself with such precision and success? The answer is because this unit had trained intensively before the engagement using SIMNET. The *73 Easting Project* was a collaborative study conducted jointly by the *Institute for Defense Analyses (IDA)*, DARPA, and the U.S. Army. With the development of a database and the use modern computer simulation technology, the soldiers were able train in a virtual re-creation of the minute-to-minute activities of each participating tank, armored vehicle, truck, and infantry team.

After the engagement more information was collected by extensive engineering surveys of the battlefield immediately after the fighting. Exhaustive participant interviews were included. This information was further integrated into the simulation of the battle for future training. The Battle of 73 Easting and the post-analysis proved the significance of computer simulation training in and of itself and with future training with its ability to test alternative cause-and-effect hypotheses with factual and counterfactual analysis. SIMNET would now include conducting controlled experiments by changing key characteristics of the historical event, then re-fighting the simulated battle and observing the effects on the presumed outcome.

Tied to the events in Iraq was the establishment of the *Executive Council on Modeling and Simulation* by the U.S. Department of Defense in 1990. In 1991 a Defense Modeling and Simulation Office was established with large investments to advance *Modular Semi-Automated Forces—ModSAF*. ModSAF is a set of software modules and applications used to construct *Advanced Distributed Simulation* and *Computer Generated Forces* applications. ModSAF modules and applications allow a single operator to create and to control large numbers of entities that are used for realistic training, testing, and evaluating on the virtual battlefield. Funding also went into the advancement of the *Joint Simulation System (JSIMS)* to develop training tools and future M&S capability, in particular simulation improvement.

Advancements in computer software and hardware as well as artificial intelligence and software agents have hastened the pace of the maturation of M&S as a discipline and tool. These additional elements that now comprise M&S enhance the capabilities of simulation for more complex phenomena such as the human personality in social and conflictual simulations. In the early 1990s, military M&S practitioners began to explore ways to link stand-alone simulations used to model and represent distinct real-world functions into a federation of simulations where simulation entities were given semi-automated behaviors. This is commonly called *semi-automated forces (SAF)*. The initial efforts to link simulations showed promise and led to standards in simulation data exchange and the establishment of protocols for creating simulation federations [9].

As semi-automated behaviors were being explored, a closer look at human behavior was underway. It was during this time that a new type of modeling was taking form—**behavioral modeling**. A behavioral model is a model of human activity in which individual or group behaviors are derived from the psychological or social aspects of humans. Behavioral models include a diversity of approaches; however, the computational approaches to human behavior modeling that are most prevalent are social network models and multi-agent systems.[6] Behavioral modeling can be used to provide qualitative analysis of a specific foreign leader, or assess the movement of civilian populations in duress, or understand how culture and religion can affect social behavior [15]. Behavioral modeling allows for the incorporation of socially dependent aspects of behavior that occur when multiple individuals are together. This type of modeling is now being used in fields of study that include observations of human behavior be they individual, group, or crowd behaviors. Education, psychology, industry, and transportation are just some examples of behavioral modeling users. (A more detailed discussion on the social sciences and behavioral modeling can be found in Chapter 8.)

The Department of Defense has also made use of the behavioral modeling. In fact, behavioral modeling research would become a significant component of military M&S. The M&S academic community and the Department of Defense analysis community had begun expanding its research to include social network analysis and crowd modeling. However, the M&S industry stayed close to military applications focusing their work on meeting the needs of the U.S. military's *Simulation, Training, and Instrumentation Command (STRICOM)* in Orlando, Florida; the *Air Force Research Laboratory* in Dayton, Ohio; the *National Simulation Center*, and the *U. S. Army Training and Doctrine Command (TRADOC)* both at Fort Leavenworth, Kansas. The *U.S. Joint Forces Command (USJFCOM)* in Suffolk, Virginia, also became very involved in M&S. In 1997 the USJFCOM partnered with Old Dominion University in establishing the Virginia Modeling, Analysis and Simulation Center—VMASC.[7]

It was not long, however, before the explosive growth of computer games for entertainment and the emergence of new uses for M&S shifted the focus of the industry. By the latter half of the decade, the companies that grew the military M&S

---

[6] A Social Network Model is a model of social behavior that takes into account relationships derived from statistical analysis of relational data. A Multi-Agent System focuses on the way in which social behavior emerges from the actions of agents [15].

[7] The Virginia Modeling, Analysis and Simulation Center (VMASC) is a multi-disciplinary modeling, simulation and visualization collaborative research center managed through the Office of Research at Old Dominion University. VMASC supports the University's Modeling and Simulation (M&S) graduate degree programs, offering multi-disciplinary M&S Masters and Ph.D. degrees to students across the Colleges of Engineering and Technology, Sciences, Education, and Business. With numerous industry, government, and academic members, VMASC furthers the development and application of modeling, simulation, and visualization as an enterprise decision-making tool and promotes economic development. http://www.vmasc.odu.edu.

industry began exploring a variety of new uses for M&S. Today, M&S can be found in just about every research and training institution or venue. M&S is being used in medical and health-care fields, logistics and transportation, manufacturing and distribution, communications, and virtual reality and gaming applications for both entertainment and education. As you will see, M&S has expanded into many application areas.

APPLICATION AREAS

Once a primary training mechanism for the military, modeling and simulation is now being used in a variety of domains to include medical modeling, emergency management, crowd modeling, transportation, game-based learning, and engineering design—to name a few. Within various forms of media, M&S has already made inroads in a number of liberal arts disciplines. With the advent of modeled and simulated historic events, television has made significant strides in portraying the Ancient World as historians have researched and perceived it to be. Advertisements for complex wireless telephone systems have employed simulated, visualized geographic data that scans large crowds then zeroes in on the individual user all in the name of selling the most extensive wireless phone service. Avatars have replaced humans for interfacing within complex communication systems. These applications of M&S have all come to pass in a somewhat indirect fashion and they are a part of our lives, although we may not recognize or even realize—this is M&S. However, as an academic tool, which is what we are endeavoring to understand, M&S has a more formal role.

M&S applications are used primarily for analysis, experimentation, and training. *Analysis* refers to the investigation of the model's behavior. *Experimentation* occurs when the behavior of the model changes under conditions that exceed the design boundaries of the model. *Training* is the development of knowledge, skills, and abilities obtained as one operates the system represented by the model. As we can see, M&S is multi-faceted and it can be used as a tool, an enabling technology. It is this property that allows us to use M&S in many disciplines. What is becoming more and more apparent to traditional producers and users of M&S is that there is a "richness of the possibilities ... and synergies with related disciplines" [16].

M&S can be applied in any field where experimentation is conducted using dynamic models. This includes all types of engineering and science studies as well as social science, business, medical, and education domains. As we have learned, the military were the first users of M&S and their applications ranged from the early models of traditional wargames to scene generation (battlefield simulation) to missile defense. The military continues to be the largest consumer of modeling and simulation; however, there has been an additional component to military decision-making via modeling and simulation that incorporates modeling the intangible aspects of a military intervention. Things like the politics of the region, the effect of war on the regional economy, and the outcome of diplomatic exchanges. M&S is often the only tool capable of solving complex problems because it allows for an understanding of system dynamics and it includes enabling technology both of which provides a means to explore credible solutions.

There are two types of modeling and simulation activity that can be distinguished depending upon whether or not the simulation program runs independently from the system it represents. **Stand-alone simulation** follows the H.A. Simon notion of *learn by doing* or train as your operate. **Integrated simulation** is used to enrich and support real systems. For many non-engineering and non-science students, Stand-alone Simulation may be best suited for your needs. Often, stand-alone application areas are grouped into five categories [9]:

- *Training*—goal is to provide real-world experience/opportunities in a controlled environment
- *Decision Support*—to provide a descriptive, explanatory, predictive tool or to provide an evaluative, prescriptive tool
- *Understanding*—this type of modeling and simulation facilitates testing a hypothesis relative to the structure and function of a complex system
- *Education and Learning*—used for teaching and learning systems with dynamic behavior and with serious gaming (this is also called game-based learning)
- *Entertainment*—simulation provides a realistic representation for elements possessing dynamic behavior

Training, Decision Support, and *Understanding* aim to provide a level of proficiency. An example of Stand-alone Simulation for *Training* is a simulation used to create an environment that focuses on game theory. Because it is focusing on game theory, the simulation may be a **zero-sum simulation** intent on honing the user's decision-making and communication skills. This is done by enveloping the user in battle simulations at different levels, or in peace operations, or in conflict management and allowing him to negotiate a solution. Another example of Stand-alone Simulation for *Training* is the **virtual simulation** with limited environmental interactions used to develop the motor skills of the user.

Whether it is *Training, Decision Support, Understanding,* or *Education and Learning*, M&S applications can be found in a number of research areas. Below are brief descriptions of four of the M&S application areas being developed at the VMASC. (The application areas introduced below will be addressed in greater detail in Chapter 8.)

Transportation M&S

Almost everyone is touched by some form of transportation every day of their lives. We rely on both public and private systems to get us to work, school, shopping, or to our favorite form of recreation. We take for granted the underlying system supporting this complex network of roadways, rail, and public transit routes. And these systems are very complex. They consist of miles of road surface and track with multiple control mechanisms to regulate the flow of vehicles. The systems have become so complex that a seemingly minor adjustment to the timing of a traffic light in one part of the city can have a drastic effect on traffic movement miles away.

Traffic engineers employ simulation to test these adjustments for just this reason. It is far better to see the results in a simulation and watch traffic back up there than it is to have hundreds of frustrated motorists wasting valuable time traveling at a speed far below their expectation. Large-scale regional traffic simulations, known as *Macroscopic Traffic Simulations*, are capable of showing the effects of these changes on very complex networks of roadways covering large regional areas.

Microscopic Traffic Simulations, those that model the individual movement of cars based on physics parameters such as velocity and acceleration, are employed to study smaller sections of roadway such as a particular intersection or a certain road corridor. These simulations provide exceptional detail to understand how individual cars will be affected if a new off ramp is added or lane configuration is changed. Using simulation for transportation planning is significant for understanding the implications of change in these very complex systems in today's environment.

Business M&S

Also known as Decision Support or Enterprise Engineering, Business M&S can be defined as a system of business endeavors within a particular business environment created to provide products and services to customers. Most enterprises are an integration of businesses, processes and infrastructure, and resources. M&S can assist in investigating, designing, and implementing solutions to complex enterprise issues such as optimally allocating scarce resources while considering stochastic (characterized by *conjecture* and *randomness*) and ill-posed environments. To address diverse complex and relevant enterprise issues it is necessary to bring together a multi-disciplinary team having expertise in operations management, operations research, industrial engineering, modeling and simulation, marketing, economics, decision science, and mathematics. Here are some of the core research areas:

1. *M&S in Manufacturing Enterprise Engineering (M&S-MEE)* addresses research on design, planning, and control of operations in manufacturing enterprises. Contributions extend the range of analytical and computational techniques addressed to these systems, and novel models offering policy knowledge of applicable solutions in manufacturing environments.

2. *M&S in Operations Research (M&S-OR)* addresses research on progress in the structures and properties of models and procedures derived from studying operations. The focus of the cluster is on researching, creating, and/or improving analytical and computational techniques while emphasizing the relevance of the work in significant applications.

3. *M&S in Service Enterprise Engineering (M&S-SEE)* addresses research on design, planning, and control of operations and processes in commercial and institutional service enterprises. As in M&S-MEE, contributions extend the range of analytical and computational methods addressed to these systems and novel models offering policy knowledge of applicable solutions. Research areas include: supply chain management, health care operations, retailing, and hospitality.

Medical M&S

M&S can assist many fields within the medical profession including training, treatment, and disease modeling. To address these problems VMASC and Old Dominion University (Norfolk, Virginia) have assembled a diverse, multi-disciplinary team having expertise in clinical medicine, modeling and simulation, engineering (mechanical, electrical, biomedical), exercise science and athletic training, human factors psychology, computer science, epidemiology, biology, mathematics, and tumor biology. The research program is a collaborative effort to develop innovative solutions to the aforementioned problems through a series of targeted research projects in core areas such as:

1. *M&S for Improved Training of Medical Professionals* Several recent studies have shown that the U.S. health care system is not as safe as it should be. It has been estimated that medical errors contribute to as many as 98,000–195,000 deaths annually in U.S. hospitals with a cost to society of $37 billion. Simulator systems for training healthcare providers have only become commercially viable within the last ten years. These systems allow trainees to learn fundamental procedures without putting patients at risk, can expose trainees to rare or unusual conditions, and reduce the need for cadavers and animal models. Unfortunately, there are large gaps between training systems currently available and the needs of educators across medical specialties. For instance, there are few systems for training in specialties such as family medicine or obstetrics and gynecology. Further, there are virtually no systems available that address the problem solving and decision-making skills of more advanced trainees. Research in Advanced Surgical Skills, Virtual Pathology, Standardized Patients, and Labor and Delivery are just a few of the areas being addressed.

2. *M&S to Improve Treatment* The treatment of disease and injury is primarily based on the experience of the physician who is treating the problem, which may result in treatment failures, unsatisfactory results, and unnecessary treatments. M&S can be used to develop new treatments, assist in decision making, and to optimize treatment. It is believed that M&S will be most beneficial in developing nonoperative and nonpharmaceutical methods to treat and prevent the progression of osteoarthritis, optimizing hardware and hardware placement for implanted devices used in orthopedic surgery, developing novel, virtual reality–based rehabilitation methods, and minimizing the dose of radiation that healthy tissue is subjected to during cancer treatment. Research in Osteoarthritis Treatment, Optimization of Orthopedic Fixation Device, and Radiation Beam and Dosage Profiling are some of the projects underway.

3. *Disease Modeling*   Multi-scale simulation models can provide opportunities to develop theories to explain the spread of disease, tumor metastasis, and the effectiveness of vaccination. Key to this is researching the dynamics and control of infectious diseases using mathematical modeling and computer simulation. Disease modeling also requires an understanding of how spatial heterogeneity impacts the spread of a given disease and the implications that brings to controlling the disease. Simulations that prescribe the optimal control techniques needed to identify the most effective disease intervention strategies are an integral part of disease modeling.

4. *M&S of Hospital Management in situations that involve Homeland Security*   Hospitals today are facing an ever-increasing demand for their services. They are at capacity or near capacity on a daily basis. But what if some type of mass casualty event like a terrorist attack or a major chemical spill should occur? These events could produce hundreds or even thousands of casualties. How will the public health system and hospitals respond? What should they have in place to support this type of disaster? Hospitals and public health officials have turned to M&S to answer these questions. Researchers have built simulations that play out these mass casualty events to see the extent to which the public is affected. They have also built simulations of hospital systems to investigate the most efficient way to treat these large volumes of patients and still provide the level of care expected of them. In short, M&S has provided significant insight to the health care industry for these catastrophic events. They are now better able to understand the implications of these events and properly prepare for them. These and other projects engaging the medical profession with M&S tools are the future of medical research and training.

Social Science M&S

For social scientists the traditional methods of modeling include statistical modeling, formal modeling, and agent-based modeling. *Statistical modeling* is the traditional method for the discovery and interpretation of patterns in large numbers of events. *Formal modeling* is a method that provides a rigorous analytic specification of the choices actors can make and how those choices interact to produce outcomes. *Agent-based simulation modeling* allows for the observation of aggregate behaviors that emerge from the interactions of large numbers of autonomous actors.

Integrating this traditional modeling and analysis capacity with other forms of modeling (simulation and visualization) serves as a tool for expanding and communicating the social scientist's grasp of the subject area being investigated and for providing a much denser schematic for the engineer's model. This relationship will accelerate interdisciplinary research efforts on the part of engineers and social scientists and it is a very good response to changing research requirements.

Every branch of the U.S. military has recognized the need to integrate historic, cultural, and political awareness into its decision-making capacity. In fact, much of the federal government research funding is coming from the analysis sector of the military and the *Department of Homeland Security (DHS)*. The social sciences are now integral to problem solving and decision-making for every aspect of governance to include the U.S. military, DHS, and Nongovernmental Organizations (NGOs). All of these institutions must react and/or respond to disturbances that are the result of insurgencies, military action, terrorist attack, and natural disaster.

Research from the Department of Defense is now requiring the incorporation of geopolitics, culture, religion, and political economy to better understand how Diplomatic, Intelligence, Military, and Economic (DIME) factors affect real-time, tactical decision-making. A growing area of research for the Department of Defense is centered on Political, Military, Economic, Security, Information, and Infrastructure (PMESII) aspects.

As we have learned, the U.S. military employs simulation to train all levels of its service personnel and to analyze complex policy and warfighting plans. Most of these simulations represent only the military aspects of warfare and ignore the political, social, and economic aspects that are vital to understanding **Effects-based Warfare** currently employed by combatant commanders.[8] Without accurate models of these social science areas combatant commanders are unable to test effects-based strategies and plans beyond tabletop exercises. These tabletop exercises do not provide the breadth of results for commanders to gain significant insight into the consequences of their decisions. Simulations that contain social science models will much better represent real-world consequences of military actions that often spill over into the political, social, and economic areas of a country or region. The Department of Defense is aware of this and, as mentioned above, is expanding its research efforts to include this type of data.

As the need for social science qualitative and quantitative analysis expands into the military and homeland security domains, a number of other equally challenging nonmilitary areas can be explored such as:

Foreign policy issue of sanctions—a system dynamics approach to measuring the effects of failed sanctions on the behavior of the civilian population, the insurgency, and/or the controlling regime

Global and national issue of energy dependency—a U.S. and/or European study of the economic, political, and social capacity to manage energy crises

---

[8] U.S. Joint Forces Command defines Effects-based Warfare as the application of armed conflict to achieve desired strategic outcomes through the effects of military force. http://www.jfcom.mil/about/glossary.htm#E.

National issue of immigration—the layered effects (labor, education, healthcare) of illegal immigrants in a specific community (state, region, or national analysis)

National education issue of game-based learning—creating educational games in the subjects of history and geography

Voting habits in countries with dubious election processes—develop a model that couples statistical data with empirical findings to address voter participation

Local government issue of urban design—continued research in modeling urban development to prescribe improvements in land use, transportation, and infrastructure

These lists are in no way exhaustive, but serve to introduce you to just some of the different M&S applications in the areas of transportation, business, medical, and social science M&S. You can be sure, however, that there are many, many uses of M&S. As you read ahead, you will come to appreciate that there are also many advantages to M&S.

USING MODELING AND SIMULATION: ADVANTAGES AND DISADVANTAGES

In 1998 the Institute of Industrial Engineers (IIE) listed the advantages and disadvantages to using modeling and simulation [1]. From their list it is easy to see why many would choose to apply M&S to research and training. Here are some of the advantages to using modeling and simulation:

- the ability to *choose correctly* by testing every aspect of a proposed change without committing additional resources
- *compress and expand time* to allow the user to speed up or slow-down behavior or phenomena to facilitate in-depth research
- *understand why* by reconstructing the scenario and examining the scenario closely by controlling the system
- *explore possibilities* in the context of policies, operating procedures, methods without disrupting the actual or real system
- *diagnose problems* by understanding the interaction among variables that make up complex systems
- *identify constraints* by reviewing delays on process, information, materials to ascertain whether or not the constraint is the effect or cause
- *develop understanding* by observing how a system operates rather than predictions about how it will operate
- *visualize the plan* with the use of animation to observe the system or organization actually operating

- *build consensus* for an objective opinion because M&S can avoid inferences
- *prepare for change* by answering the "what if" in the design or modification of the system
- *invest wisely* because a simulated study costs much less than the cost of changing or modifying a system
- *better training* can be done less expensively and with less disruption than on-the-job training
- *specify requirements* for a system design that can be modified to reach the desired goal

It is obvious that there are many uses and many advantages to using M&S. The IIE also made note of some of the disadvantages to using M&S. The list is noticeably shorter and it includes things such as the *special training* needed for building models; the *difficulty in interpreting results* when the observation may be the result of system inter-relationships or randomness; *cost in money and time* due to the fact that simulation modeling and analysis can be time consuming and expensive; *inappropriate use* of modeling and simulation when an analytical solution is best.

CONCLUSION

Let's review what has been introduced in this chapter. *Models* are approximations of events, real events as in case studies, or contrived events as in use-case studies. We create models from data so our research of the event or details that go into the use-case must be accurate to ensure that the model is sound. With a reliable model we can develop a hypothesis or research question that requires our observation of the model. We observe the model via *simulation*, and as we have learned we can modify and repeat the simulation. Often, models include *systems* or collections of different elements that together produce results not obtainable to the elements alone. We then conduct our *analysis* of the simulations to draw our conclusion or verify and validate the research. The ability to apply visualization allows us to communicate or present the model, the simulation, and the conclusions we have drawn. All of this is *learning by doing*.

As we have seen the *history of M&S* is quite lengthy, especially if we start with the modeled battlefields and wargames of the Ancient World. Let's not forget the mini theatre of operations that the game of chess models. Scientists like Leonardo DaVinci modeled everything from government edifices, to life-size busts, to bridges. He also tested his inventions and, in doing, so left us with the first systematic explanation of how machines (or systems) operate.

The military no doubt engaged M&S to the fullest during the 19th and 20th centuries. By the 1990s *behavioral modeling* was integrated into military applications of M&S. Behavioral modeling is also present in many other research fields because it focuses on human activity and behavior derived from the psychological and social aspects of humans.

We were also introduced to a few of the many applications of M&S such as transportation, business, medical, and the social sciences. The many advantages to conducting research with the use of M&S are also noted. Some of these advantages are exploring possibilities, diagnosing problems, visualizing a plan, training, and specifying requirements.

The multidisciplinary approach to the principles of M&S that are outlined in the following chapters will help you understand the theoretical underpinnings of M&S, explore M&S practical domains, and observe real-world M&S applications. As with many other students you, too, may decide that M&S education and research will be your discipline of choice.

KEY TERMS

modeling	Military-Industrial	stand-alone simulation
simulation	Complex	integrated simulation
visualization	avatars	zero-sum simulation
analysis	behavior modeling	virtual simulation
system		

FURTHER READING

CLARFIELD GH. *Security and Solvency: Dwight D. Eisenhower and the Shaping of the American Military Establishment*. New York: Greenwood Press, 1999.

LANGLEY P, SIMON HA, BRADSHAW GL, and ZYTKOW JM. *Scientific Discovery: Computational Explorations of the Creative Processes*. Cambridge, MA: The MIT Press, 1987.

ROMEI F. *Leonardo da Vinci: Artist, Inventor and Scientist of the Renaissance*. New York: Peter Bedrick Books, 1994.

SHENK D. *The Immortal Game: A History of Chess, or How 32 Carved Pieces on a Board Illuminated Our Understanding of War, Art, Science and the Human Brain*. New York: Doubleday Press, 2006.

VERNON A, HOLMES R, DOWNEY G, TRYBULA D, CREIGHTON N. *The Eyes of Orion: Five Tank Lieutenants in the Persian Gulf War*. Kent, OH: Kent State University Press, 1999.

REFERENCES

1. COLWELL RR. Complexity and Connectivity: A New Cartography for Science and Engineering. Remarks from the American Geophysical Union's fall meeting. San Francisco; 1999.
2. Simulation-Based Engineering Science: Final Report, May 2006. Available at http://www.nsf.gov/pubs/reports/sbes_final_report.pdf. Accessed 2008 Jan 2.
3. BANKS J, editor. *Handbook of Simulation: Principles, Methodology, Advances, Applications, and Practice*. New York: John Wiley & Sons, Inc; 1998.
4. FISHWICK PA. *Simulation Model Design and Execution: Building Digital Worlds*. Upper Saddle River, NJ: Prentice Hall; 1995.
5. LAW AM, KELTON WD. *Simulation, Modeling, and Analysis*, 4th ed. New York: McGraw-Hill, Inc; 2006.
6. IBM Research—Deep Blue. Available at http://www.research.ibm.com/deepblue/watch/html/c.shtml. Accessed 2008 Jan 2.

7. YOUNG JP. *History and Bibliography of War Gaming.* Department of the Army; Bethesda, Maryland 1957.

8. Link Simulation & Training. Available at http://www.link.com/history.html. Accessed 2008 Jan 2.

9. ÖREN TI. (2005—Invited Tutorial). Toward the Body of Knowledge of Modeling and Simulation (M&SBOK). In: Proceedings. of I/ITSEC (Interservice/Industry Training, Simulation Conference); 2005 Nov 28–Dec 1; Orlando, FL. pp. 1–19.

10. President Dwight D. Eisenhower's Farewell Address (1961). Available at http://www.ourdocuments. gov/. Accessed 2008 Jan 2.

11. Spacewar. Available at http://www/wheels.org/spacewar. Accessed 2008 Jan 2.

12. Mazewar. Available at http://www.digibarn.com/history/04-VCF7-MazeWar/index.html. Accessed 2008 Jan 2.

13. TACDEW. Available at http://www.ntsc.navy.mil/Programs/TrainerDescriptions/Surface/TACDEW. cfm. Accessed 2008 Jan 2.

14. SIMNET. Available at http://www.peostri.army.mil/PRODUCTS/PC_BASED_TECH/. Accessed 2008 Jan 2.

15. CARLEY K. Social Behavior Modeling. In *Defense Modeling, Simulation, and Analysis: Meeting the Challenge.* Washington, DC: National Academies Press; 2006.

16. ÖREN TI. Maturing Phase of the Modeling and Simulation Discipline. In: Proceedings of Asian Simulation Conference 2005 (The Sixth International Conference on System Simulation and Scientific Computing (ICSC'2005); 2005 October 24–27; Beijing P.R. China. Beijing P.R. China: International Academic Publishers—World Publishing Corporation; 2005. pp. 72–85.

Chapter 2

The Role of Modeling and Simulation

Paul F. Reynolds, Jr.

INTRODUCTION

Simulation is transforming the way researchers think and explore and the way all of us live. Simulation is now considered a third and extremely useful avenue of scientific pursuit, along with theory and experimentation. It is used to gather insight, validate models and experiments, train humans, support statistical analyses, drive computer animations, control real-time processes, predict potential outcomes, test and evaluate new systems, and support what-if analyses, among others. It is performed on computing platforms ranging from desktop computers to the fastest high-performance parallel computers and even large-scale geographically distributed systems. Armed with a broader range of studies, students, analysts, and decision makers gain insight, become better informed, and reduce risk in their analyses and decisions. Analysts and decision makers use simulation routinely to support their analyses, predictions, and decisions. One cannot pass through a day without encountering or at least benefiting from the impact of modern uses of simulation.

Simulation has a spectrum of usage patterns ranging from problem solving to insight gathering under conditions of high uncertainty. Points along this spectrum are defined by how much insight gathering is occurring and how much uncertainty there is about the correctness of the simulation and the understanding of the phenomenon being simulated. A simulation used in problem solving is generally believed to be a proper representation of the phenomenon it was designed to simulate. At the problem solving end of the spectrum, the simulation is a trusted tool. A simulation used in gathering insight is, by contrast, more likely to be seen as an incomplete and possibly incorrect model of the phenomenon that is being studied. Because critical

Principles of Modeling and Simulation: A Multidisciplinary Approach, Edited by John A. Sokolowski and Catherine M. Banks.

information about the phenomenon being simulated may be lacking, the simulation is not assumed to be a proper representation of it. In other words, problem-solving simulations have little to no uncertainty associated with them, while insight-gathering simulations may have a great deal of uncertainty associated with them. As a result, users of problem-solving simulations generally trust them and don't feel a need to change them, while insight gatherers see their simulations as a part of a learning process where the simulation itself will have to be adapted to meet new requirements as insight is gained from previous incarnations and uses of the simulation. A simulation used for insight gathering may ultimately be thrown away once its user decides all the useful insight to be gained has been learned. Problem-solving simulations may stay in use for years.

In the following sections we discuss the use of simulations both in problem-solving and in gathering insight. Because uncertainty plays such a significant role in differentiating problem-solving usage from insight-gathering usage we also explore uncertainty and its effects before discussing simulation used for insight.

USING SIMULATIONS TO SOLVE PROBLEMS

In a problem-solving scenario a simulation is considered to be valid for the context in which it is being applied. For example, if a researcher is studying the trajectories and impact points of munitions using a simulation meant to support that study, then it is assumed that the simulation is used for circumstances that match those expected by the designer of the simulation. For example, the designer may have assumed that the platform from which the munition is fired is stationary. The user of the simulation should use it only under similar assumptions. Otherwise the simulation can no longer be trusted to produce meaningful results.

Simulations used for problem solving generally have parameters associated with them, to allow the user to define selected portions of the context for the simulation, to choose important controls in the simulation, and perhaps to include data that help to define the way the simulation should operate. For the ballistics simulation, contextual parameters may include weather conditions, position of the platform on 3D terrain, and parameters for probabilistic processes used to represent complex processes in the phenomenon of interest. These probabilistic processes are often used by designers as a way of addressing aleatory uncertainty, which we discuss further in the next section. Control information could include how the ballistics computations are to be performed, and what numerical accuracy should be used. Simulation input data could include how often the weapon is fired (can affect ballistics), what the muzzle angle is, and similar parameters a user of the weapon being simulated may be able to set.

Examples of problem-solving simulations include ballistics computations (one of the first uses of simulation, in the 1940s, on modern computing technologies), automobile stability under adverse driving conditions, outcome of coin tossing, tomorrow's weather prediction, commercial nuclear waste storage in Yucca Mountain, disease spread, impact of various cache architectures on processor performance, and observable physical motions of a bicyclist following a prescribed trajectory at a given velocity. There's an infinite set of possible problem-solving applications for simulation, so this list is only a sampling. However, it provides some insight into the kinds of applications that problem solvers using simulation may be trying to study. In general, if a user is asking:

What would happen if ...?

How will a ...?

Why would a ...?

Can a ...?

Does the ...?

Should we ...?

and using a simulation to provide the answer, then that user is typically performing problem solving with the simulation. These are not the only questions that could be asked when using a simulation for problem solving, but they reflect the kinds of questions that would be asked. Consider the application examples listed in the preceding paragraph. When used in problem solving, the following kinds of questions might be asked:

Ballistics

What: What firing velocity would be required to make a munition with certain physical characteristics damage the weapon muzzle?

How: How would a munition with certain physical characteristics and fired at a selected velocity pass through the air?

Why: Why would a munition with certain physical characteristics and fired at a given velocity fly off course?

Can: Can a munition with certain physical characteristics and fired at a given velocity travel more than X kilometers?

Does: Does the ballistics simulation output match the answers provided by mathematical ballistics models?

Should: Should we use a munition with certain physical characteristics to perform a given task?

Coin Tossing

What: What are the most likely outcomes (sequences of heads and tails) of flipping a fair coin 1000 times?

How: How is it possible to observe 900 heads and 100 tails in 1000 flips of a fair coin?

Why: Why in repeated sessions of 1000 flips of a fair coin is the outcome 900 heads and 100 tails rarely observed?

Can: Can 900 heads and 100 tails be observed in 1000 flips of a fair coin with a probability greater that 0.1%?

Does: Does the simulated distribution of possible outcomes for flipping a fair coin match the expected distribution of outcomes?

Should: Should we count on 1000 flips of a fair coin coming up heads 900 times and tails 100 times to not occur?

(This coin example provides an excellent opportunity to make the point that whenever simulation is considered for problem solving, you must be sure to consider analytic methods as an alternative first. All six coin-tossing questions can be answered in minutes by anyone possessing an easily acquired knowledge of combinatorics. The weeks it could take to answer the five questions above using simulation is time lost!)

Commercial Nuclear Waste Storage in Yucca Mountain

What: What amounts of rainfall would have to occur to make dangerous amounts of nuclear waste materials leach out of Yucca Mountain?

How: How could it come to pass that dangerous amounts of nuclear waste materials leach out of Yucca Mountain?

Why: Why do nuclear waste storage containers made of Alloy 22 not corrode in an environment similar to Yucca Mountain's?

Can: Can an aluminum outer container for waste storage prevent corrosion under expected environmental conditions?

Does: Does the simulated rate of corrosion for an Alloy 22 container match known experimental results?

Should: Should America's commercial nuclear waste be stored in Yucca Mountain? (Learn more about this question by exploring the "Total Systems Performance Assessment" study—TSPA—meant to answer this very question.)

We leave it to the reader to demonstrate six questions of a similar nature for automobile stability analysis, weather prediction, cache performance analysis, and other problem-solving applications listed above.

At the beginning of this chapter we listed a number of uses for simulations. They include gathering insight, validating models and experiments, training humans, supporting statistical analyses, driving computer animations, controling real-time processes, predicting potential outcomes, testing new systems, supporting what-if analyses, and enhancing teaching and education at all levels. All of the uses listed tend to fall under the role of using simulations for problem solving. Because we have distinguished the problem-solving and insight-gathering roles of simulation earlier in this chapter, the reader may be a little confused. Fear not. There is a distinction based on who is gathering the insight, and the degree of uncertainty present in the simulation. Problem-solving users can gain insight from a simulation that is trusted in a manner that is distinct from the kind of insight gathered by a simulation developer as s/he attempts to build a valid, trusted simulation.

Gathering Insight

Any time the user of a simulation deemed valid for his or her use is using the simulation for problem solving, there will likely be some increase in the user's understanding of the modeled phenomenon. This is generally an incidental acquisition of insight. The user's intent will likely be one of the problem-solving activities described in subsequent paragraphs. Different users may gain different kinds and levels of insight. Generally successful solution of the problem the user means to address will not depend on the insight the user gains. Even when it does, the insight gained is not for the purpose of improving the simulation. We distinguish this user gathering of insight from a more purposeful acquisition the designer conducts as described in a bit later.

Validate Models and Experiments

Often, researchers will come up with analytic models for the behavior of a phenomenon of interest. Also, if the phenomenon of interest can be studied directly (e.g., corrosion rates of stainless steel containers and human reactions to carefully crafted social situations can be studied experimentally under controlled conditions, but collisions of galaxies cannot), the researcher may have gathered data from instrumentable experiments. For both analytic models and experimental data the researcher may turn to simulation as a method for validating the model or data. Consider a model first. As an example, there is a simply stated analytic result in classic queuing theory that says the expected length of a queue (e.g., waiting in line at your bank) is always equal to the average rate at which the queue's server (e.g., the bank teller) can process customers times the expected amount of time each customer will ultimately have to wait in the queue. This is known as Little's Law and is expressed concisely as:

$$L = \lambda W$$

where L is the mean queue length, λ is the mean service rate (customers processed per unit of time), and W is the mean wait time. For this or any analytic model, a researcher may choose to implement a simulation that models the phenomenon of interest, perhaps in significant detail, and then compare the results with the results the model gives. In the case of Little's Law there is a formal mathematical proof of its correctness, so a simulation that validates it is unnecessary. However, for most analytic models no proofs of correctness will be found, and so simulation may become the best avenue for validating the analytic model.

In the case of Little's Law, if the researcher chooses to use simulation to validate the analytic model, then it could be done in about a page of code in a typical programming language. The user would simulate the arrival of customers, probably using a randomly sampled distribution for determining arrival times of customers, and another distribution for determining times required for individual customers to be serviced by the server (teller). Coupled with some fairly straight-forward statistics gathering about each of the customers entering and exiting the queue, the researcher could produce results about mean queue length and mean wait time and, along with the simulated mean customer service rate, establish whether Little's Law agrees, and is thus deemed valid.

A similar approach can be taken to validate data collected from a set of experiments. Often, when one gathers data from experiments there will be variations in results because of measurement errors, simple variance in the data collected, or possibly incorrect setup in the experiments themselves. A simulation that the researcher believes accurately reflects the expected behavior of the phenomenon being studied can be constructed in order to test whether the collected data matches corresponding simulated results within an acceptable margin of error. If it does, then the data may be declared to be representative of the phenomenon modeled by the simulation.

Both model validation and experimental data validation are addressed with a "Does a ...?" question. The question is, "Does the trusted simulation match the analytic model (alternatively, the collected data from experiments)?" An acceptable match implies validity.

Note researchers also use data collected from experiments to validate simulations (the converse of what we've just discussed). This is a different process with different objectives and is discussed in the section on the subject of gaining insight.

Training Humans

Simulation is used extensively to drive training exercises. Soldiers, pilots, operators of dangerous equipment, and others have used simulation-based training for years. This use of simulation also reflects problem solving. In this case the simulation is considered valid and the training exercise becomes a "Can the ...?" or "Does the ...?" experiment where the relevant question is "Can the trainee perform function X?" or "Does the trainee know how to respond to condition Y?" Note that training is very similar to validating experimental data if we regard the trainee's responses as data. In both cases the simulation is used to measure whether the data is within an acceptable margin of error. In the case of a trainee, responses outside of the acceptable limits means the training has not yet succeeded.

Support Statistical Analyses

Once a simulation is trusted—deemed valid—it can be used to conduct experiments that yield useful statistics about possible outcomes. A simulation can produce different results on different executions because its input parameters are changed, or because it contains random ("stochastic") processes, or both. A researcher will typically set up a set of experiments, knowing where input or internal to the simulation conditions can change, and then initiate a set of simulation executions that produce a set of results. That set of results can then be analyzed for properties of interest, such as means and variances associated with simulation outputs of interest. Considered as a form of problem solving, statistical analyses answer largely "How can ...?" questions, namely, "How can the outputs of interest vary given a prescribed set of changes in simulation inputs?" Statistical analysis is one of the problem-solving activities where the user may very well increase his or her insight into interesting behaviors of the simulated phenomenon. As mentioned previously we distinguish this user acquisition of insight from the more purposeful insight gathering activity the designer of a simulation would conduct. We will discuss this topic further the next section.

Drive Computer Animations

Simulations are often used as the engine that creates action in computer animations. Typically the simulation behind an animation is a set of equations that are evaluated numerically repeatedly. Parameters to the equations are often derived from simulation parameters or scripts meant to determine what the animation does. In this case the problem being solved is "What would the animated entity look like (the animation) when simulated using the input parameters meant to determine behavior?"

Control Real-Time Processes

In some interesting cases (e.g., oil well drilling) simulation is used in conjunction with a real-time process to predict what might happen in the real-time process in the near future, and how the real-time process might be altered to perform its job better. This kind of synergistic mating of simulation and real-time process has received a lot of attention from researchers in recent years and was the impetus for a major National Science Foundation program in the early part of this century. In the synergistic relationship between simulation and real-time process, the simulation controls the real-time process and the real-time process provides data back to the simulation so that it can better perform its analysis and control functions. In this scenario the simulation is addressing: "What would happen if ..." given its internal model of the real-time process and the data the real-time process is providing. After answering the question the simulation provides new control inputs to the real-time process and the cycle repeats.

Predict Potential Outcomes

Simulation is used extensively in weather modeling and forecasting, using a method called ensemble modeling. The weather system is generally treated as a chaotic system, meaning, in simple terms, that large changes in a weather forecast can occur with small perturbations in input conditions. Not all simulations used for prediction suffer from this output sensitivity to input conditions. Part of using a model for prediction is developing a solid understanding of how sensitive it is to changes in its inputs. Once the degree of sensitivity is established, an approach leading to effective use of the simulation can be established, e.g., ensemble modeling. Ensemble modeling as used by weather forecasters involves gathering statistical data about possible outcomes given variations in inputs. Once the statistical data is collected, there are methods for formulating a prediction with reasonable confidence, although in the case of current weather forecasting methods the predictions often suffer high levels of inaccuracy beyond about three days. Using models to predict is an attempt to answer "What would happen if ...?" where the "if" part of the question is determined by current (possibly assumed) conditions, for example, in weather modeling, current weather and atmospheric conditions.

Prediction in support of decision making is a controversial use of simulation. Because prediction often entails "What would happen if ...?" questions, and to answer them we need to assume validity of the simulation, validity of answers depends on validity of our assumption that the simulation is valid. On the other hand, simulation plays a substantial role in prediction because 1) it supports representation of conditions that may not be representable in other analysis methods, 2) it supports analysis of large numbers of cases, and 3) it frequently enables insight into behaviors related to the phenomenon under study that may not be observable in other methods of study.

Prediction is a controversial use of simulation because of lingering uncertainty about simulation validity. There are those who believe that simulations should never be used to predict, but only to provide humans with insight, so that the humans can formulate their own predictions. The basis for the never-use-to-predict argument is that humans possess a power to reason that modern digital computers have never demonstrated. So long as that is true, the argument goes, humans should do the predicting. A counter-argument is that computers can analyze more cases than humans can comprehend, and the outcome may be machine superiority. Supporting this argument in recent years are the accomplishments of modern computers with respect to chess and checkers and other competitive activities, where machines win on the basis of the large number of cases that can be analyzed. We won't resolve this argument here and now, but the reader should be aware that genuine disagreement exists, and care should be exercised when using simulations in a problem-solving manner to answer "How would a ..." questions and basing predictions solely on the simulation outcomes.

Test and Evaluate New Systems

Simulations can be used to provide inputs to actual processes, physical or logical, for the purpose of testing them. An example is software systems. When someone develops a new piece of software, simulation can be used to generate conditions meant to test the ability of the software to meet its requirements. Simulations can also be used to support evaluation of processes. A "What would happen if ..." question meant to support process evaluation can be carried out by presenting simulation outputs to the tested process that allow a user to determine whether the process responds correctly.

Support What-if Analyses

Consider a physical system that is deemed safe to explore. For those who don't read the instruction manual, a typical avenue of approach is to start pushing buttons and spinning dials, as a child might do with a new toy. Simulations can be used in a similar manner, especially since most simulations don't come with instruction manuals! The range of questions that might be asked includes all of the problem solving questions: What would happen if ...? How would/will a ...? Why would a ...? Can a ...? Does the ...? and Should we ...? As with training and statistical analysis, what-if analyses can be considered a form of insight gathering for the user. Because the user is not questioning the validity of the simulation and is gathering trusted insight about the modeled phenomenon, we regard what-if analysis as a form of problem solving.

Enhance Teaching and Education

Use of simulations in educational settings is expanding rapidly. Teachers use them because they can bring many processes and experiences to life. Students use them as an alternative method for learning. Just about all of the uses presented above, for example, supporting statistical analyses, predicting outcomes, and conducting what-if analyses, occur when simulations are used in the classroom.

Finally, we introduce the notion of **model calibration**. (Here "model" is used interchangeably with "simulation.") When the development of a simulation reaches the point that its designer believes it properly reflects the phenomenon to be simulated, the simulation may need to be calibrated before use. For example, in this century there have been a number of models developed that attempt to predict the spread of serious infectious diseases (e.g., smallpox, bird flu) throughout a population, often a large American city. These epidemiology, or disease spread, studies are designed to address questions such as "What intervention strategy should be used?" (In epidemiology models intervention strategies include vaccination and population isolation.) Before using an epidemiology simulation the developer will calibrate it on a dataset. (In the case of smallpox models these datasets are often old and for non-American populations, which raises interesting validity questions.)

Model calibration is performed in order to determine reasonable values for critical simulation parameters. In epidemiology models one of the most important parameters is "R," which is a measure of how aggressively a disease spreads from one member of a population to another. Once a model is calibrated, it is then applied to conditions for which no data exist (e.g., spread of smallpox in Portland in the 21st century). Model calibration is an important step that generally occurs between the time a user is confident the simulation is structurally correct (all uncertainty of import to the user is removed) and when the simulation is applied in untested conditions. Not all simulations need to be calibrated, but calibration does occur often. Because calibration is generally applied to simulations considered structurally valid, we regard it as a part of the problem-solving role of simulations.

If a user is uncertain about any aspect of the phenomenon they mean to be simulating, or about the validity of the simulation they've developed to represent a phenomenon of interest, then that user will not generally be using the simulation for problem solving. Rather the simulation will be used to gather insight into the phenomenon or into simulation validity. Uncertainty plays a significant role in insight gathering. We discuss the nature of uncertainty and its influences next.

UNCERTAINTY AND ITS EFFECTS

Uncertainty exists in most everything we think about and do. In everyday life there are elements of uncertainty that drive our behavior: the unreceived phone call, insufficient detail about another's expectations, when the bus will arrive. Uncertainty is similarly pervasive in modeling and simulation. When a particular physical or logical phenomenon is modeled there are typically many attributes associated with the model that just aren't understood clearly, or that are too difficult to comprehend. As this book is written, theoretical physicists are exploring the concept of string theory as a kind of unifying theory for all of physics. If string theory is ultimately shown to have a sound basis then that could have a significant impact on future theory and modeling of astrophysics phenomena, for example. Any scientist engaged in astrophysical modeling doesn't currently know what impact string theory should have on models because the nature, role, and validity of string theory is still being explored.

To understand the impact of uncertainty on model and simulation development, consider the following. According to a February 2006 report of the National Science Foundation Blue Ribbon Panel on Simulation-Based Engineering Science

(SBES): "The development of reliable methodologies—algorithms, data acquisition and management procedures, software, and theory—for quantifying uncertainty in computer predictions stands as one of the most important and daunting challenges in advancing SBES." Uncertainty's daunting nature is evident in the results of epidemiology studies conducted this century. Epidemiologists have addressed the question of government-level actions and reactions regarding the spread of infectious diseases such as smallpox and bird flu. Should a comprehensive vaccination program be initiated? How and to what degree should infected individuals be isolated, and for how long? The range of simulationists' answers to these questions is broad and full of conflict. Recently it has shown analytically that just four of the potentially hundreds of critical independent variables in these studies induce extreme sensitivity in model predictions, leading to serious conflict regarding remedial approaches involving billions of dollars and millions of people. (The interested reader is invited to consult the Elderd et al. reference listed at the end of this chapter.)

Most theoreticians who study uncertainty believe there are two kinds: epistemic (roughly "structure") and aleatory (roughly "random variation"). To better appreciate these kinds of uncertainty, consider the coin tossing example. If a fair coin is flipped 1000 times, the probability of various outcomes, for example, 402 heads and 598 tails, is well known and can be predicted using a binomial distribution. The most likely outcome is that there will be a mix of 500 heads and 500 tails. Slightly less likely is that there will be 499 heads and 501 tails, and of equal likelihood to that event is 501 heads and 499 tails. There is even a non-zero (but small) probability that there will be zero heads and 500 tails (and equally likely is 500 heads and zero tails). Now consider a smart coin, one that is biased towards coming up the opposite of what it came up the last time it was flipped. The more biased the coin is, the more likely the outcome of 1000 flips would be exactly 500 heads and 500 tails. In the case of perfect bias (always the opposite on the next flip) the only possible outcome is 500 heads and 500 tails.

Now imagine trying to simulate a fair coin. Often this is done by using a pseudo-random number generator that returns all values between zero and one with equal probability. (Thus representing what is known as a uniform random distribution over the interval zero to one.) A simulation for the fair coin could be written that interpreted any value between zero and ½ as an outcome of heads and any value greater than ½ and less than or equal to one as an outcome of tails. If you placed this simulated coin flipper in a loop and simulated flipping a coin 1000 times, the likelihoods of heads/tails outcomes would match those seen with an actual fair coin, as represented by a binomial distribution. If someone covertly replaced your fair coin with a smart coin, your simulation would not necessarily be a proper representation of the potential outcomes from flipping the smart coin 1000 times. (It would be a proper representation only if the smart coin had no bias.) In the case where the smart coin had perfect bias for coming up the opposite on the next flip, your simulation would be a terrible model of the possible outcomes, because in 1000 flips the coin could only come up 500 heads and 500 tails.

Aleatory Uncertainty (Random Variations)

There are certain phenomena and events in any environment where we have to consider them as random because we simply have no better way of characterizing them. In the coin flipping example, phenomena such as random air currents, irregularities in the surface a landing coin strikes, and forces the coin flipper places on the coin when it's flipped all would generally be treated as random. They would either be simulated using some sort of probability distribution that measurements or expert opinions suggest, or they would be ignored. In the case of the coin simulator we chose to ignore them. Whether ignored, or simulated using a probabilistic (stochastic) process, they represent **aleatory uncertainty**.

Epistemic Uncertainty (Structure Uncertainty)

If someone handed you a coin and said, "Here, simulate the possible outcomes of 1000 flips of this coin," you would be faced with a degree of **epistemic uncertainty**. Is the coin fair? Is it a smart coin? Is it a new kind of coin? If you could study many flips of the coin you might be able to transform your epistemic uncertainty to aleatory uncertainty by gaining a degree of insight and confidence about the coin's behavior—at least enough to represent it using a stochastic process. But sometimes such measurements are difficult or impossible to acquire. In that case you must deal with your epistemic uncertainty by making educated guesses about how best to simulate the behavior of the coin you were handed. You would be addressing your epistemic uncertainty.

Aleatory uncertainty can occur with any kind of simulation—those used for problem solving as well as those used for gaining insight. The presence of any significant degree of epistemic uncertainty almost always suggests that the simulation will be used for gaining insight. All of the questions associated with using a simulation for problem solving, as discussed in the preceding section, are based on the assumption that the user believes the simulation properly reflects the expected behavior of the simulated phenomenon. That is, no epistemic uncertainty that would significantly bias simulation behavior is present. If the user knows there is epistemic uncertainty associated with the simulation then that uncertainty needs to be addressed. There are basically two options for doing so: 1) Identify ways to reduce the epistemic uncertainty, or 2) Accept the risks associated with leaving the simulation as it is. Removal of the uncertainty can involve a range of activities including gathering insight into the epistemic uncertainty using the simulation as a source of data, conducting experiments to gather data and insight, or engaging in theoretical analyses that may produce insight, or any mix of these activities. Throughout this process the simulation will be adapted to reflect the user's newly acquired insight into what needs to be done to remove the epistemic uncertainty.

Accepting the risks associated with epistemic uncertainty is an alternative. On occasion the user may see no path to resolving the uncertainty, or its resolution may require too high an investment of time or other resources. Also, the user may have sufficient understanding of the phenomenon of interest to conclude that the epistemic uncertainty is not significant, and that the intended uses of the simulation will not be materially affected by its existence.

Those times when a user does elect to use the simulation to gather insight, the process leading to an acceptable version of the simulation may consume a significant amount of time, not uncommonly months or even years, and sometimes an entire professional career. We discuss insight gathering next.

GAINING INSIGHT

The second use of simulation is in gaining insight. Recall that we reserve the term **gaining insight** for acquiring an understanding of the phenomenon of interest or of how to simulate the phenomenon. Gaining insight, so defined, is the task of the simulation designer. The simulation user, as distinct from the designer, may also gain insight from the simulation, but the user's insight is often incidental and not generally associated with acquiring an understanding of the simulation or of the phenomenon of interest for the purpose of improving the simulation. Thus in this section we will look at simulation from the point of view of the designer. The designer is typically a scientist or other researcher who seeks to understand some phenomenon and capture this understanding in the form of a simulation. In the process of gaining insight about a simulated phenomenon there are certain kinds of questions being asked: In general, if a designer is asking:

What has the greatest influence?

How will X and Y interact?

Is there a way to make X happen?

Why has unexpected behavior X occurred?

What new behaviors might emerge?

and using a simulation to provide the answer, then that designer is almost always attempting to gain insight, and to reduce epistemic uncertainty. These are not the only questions that could be asked when using a simulation for gaining insight, but they reflect the kinds of questions that would be asked. Consider two of the problems introduced previously: coin tossing and the question of whether to store America's commercial nuclear waste in Yucca Mountain. We demonstrate instances of the questions listed above with respect to these two problems. For all of the questions posed, the objective of the designer may be to advance establishment of the validity of the simulation, and it may be to gain further insight into the simulated phenomenon by temporarily assuming validity of the simulation and then evaluating a simulated behavior for reasonableness. Failure of the reasonableness test would indicate that the modeled phenomenon needs to be understood better and then that better understanding needs to be captured in the simulation.

Coin Tossing

What: What has the greatest influence on the outcome of flipping a simulated fair coin? In the case of validating the simulation a designer would expect the answer to be "nothing" because the outcome of a coin flip should be completely independent of the outcomes of any previous flips. In the case of gaining further insight into the simulated phenomenon, the answer may help the designer discover the notion of independence in fair coin flipping.

How: How will the the ith and $I + 1^{st}$ coin flips interact? A designer seeking to validate the coin flipping simulation for a fair coin would expect the answer to be "In no way" because coin flips are regarded as independent events. Any other outcome requires modification of the simulation. In the case of gaining further insight into the simulated phenomenon, the outcome of the simulated coin flips may give the designer insight into how fair coins should behave. If the designer believes that the longer a coin comes up one result, the more likely it is to come up the other on a subsequent flip (a common misperception), the outcome of the simulation may lead to the discovery of independence of flips.

Is: Is there a way for two heads to come up in a row as a simulated coin is repeatedly flipped? If after repeated trials using the simulation the designer never observes two heads in sequence s/he may conclude the simulation is not yet valid because a fair coin should occasionally come up heads two or more times in sequence. For the designer seeking further insight into possible outcomes for flipping an actual fair coin, the outcome may provide insight into the fact that a fair coin will not have a smart coin's bias.

Why: Why in one trial of 1000 simulated flips did 850 heads and 150 tails occur? The designer seeking insight for validating a simulation may run many more 1000-flip trials and observe that the distribution of outcomes approximates a binomial distribution. In learning this, the designer may conclude that the simulation of a fair coin is valid. As with the "Is ..." question above, for the designer seeking further insight into possible outcomes for flipping an actual fair coin, the outcome may provide insight into the fact that a fair coin will not have a smart coin's bias.

What: What new behaviors might emerge? This is a sweeping question that may require many executions of the simulation to answer to the designer's satisfaction. For validation of the simulation, the designer would expect that no surprises would occur.

New behaviors could all be explained. For the designer seeking further insight into the phenomenon of interest, this goal is probably to learn as much about the phenomenon of interest as possible. As in the other cases above, a simulated behavior that does not agree with the designer's view of the phenomenon of interest may lead to modifications of the simulation or to a revised assessment of the phenomenon of interest.

In the case of coin flipping and problem solving we noted that simulation was not really necessary because a beginner's understanding of combinatorics is sufficient to answer all of the problem solving questions analytically. For the insight questions just discussed, combinatorics might be useful in some respects but used alone are inadequate to answer the questions posed. Unlike the problem-solving user, the insight-seeking designer is attempting to resolve epistemic uncertainty—uncertainty about the phenomenon of interest or uncertainty about the simulation meant to capture the phenomenon of interest, or both.

In the following discussion of insight seeking questions for the Yucca Mountain study, we discuss simulation validation questions only. Use of the questions for exploration of the phenomenon of interest is left to the reader.

Commercial Nuclear Waste Storage in Yucca Mountain

What: What has the greatest influence on the expected lifespan of a storage container for nuclear waste material? Asked as a part of validating a storage container simulation, this question can provide insight into the simulation's validity. For example, if the designer knows from experimentation and practice that the alloy used to make the container is the most important factor then the simulation should support this observation. What the simulation predicts provides insight into the simulation's validity.

How: How will the alloy used to make the container interact with the expected environment inside Yucca Mountain? Again, asked as a part of validating a storage container simulation, this question can provide insight into the simulation's validity.

Is: Is there a way for human exposure to unacceptable levels of radiation to occur anytime in the next one million years? Scientists know that exceptional amounts of rainfall could lead to unacceptable radiation exposure. A valid simulation would expose this result when given the right initial conditions. A designer seeking insight about the validity of the simulation could learn about its validity by creating the exceptional rainfall conditions in the simulation.

Why: Why does the simulation predict storage container failure when moisture levels and temperature values exceed certain thresholds? In pursuing this question the designer could gain insight into simulation validity by observing whether the answer to this why question matches theory and experimental results.

What: What new behaviors might emerge? As with coin flipping or any other simulation analysis, this is a sweeping question that may require many executions of the simulation to answer to the designer's satisfaction. For validation of the simulation, the designer would expect that no surprises would occur and new behaviors could all be explained.

We previously learned that problem solving simulations accept parameters to define selected portions of the context for the simulation, to choose important controls in the simulation, and perhaps to include data that help to define the way the simulation should operate. Simulations used for gaining insight accept similar kinds of parameters. Some of the context parameters may be data for probabilistic (stochastic) processes meant to address aleatory uncertainty. For example, in the coin-tossing problem, the simulation may represent air currents as a process that has well-defined but random effects on the spinning of a tossed coin. Parameters to this process could include the mean and variance for a probabilistic distribution meant to represent important properties of air currents. A developer seeking insight may vary these kinds of parameters to gain insight into how well the probabilistic process represents what the designer knows about air currents. If the process is found to not represent air currents well, then the designer may change its parameters, or replace the process completely with another. In either case the simulation is being adapted to better reflect the phenomenon of interest (coin tossing in the presence of air currents).

How might a designer establish whether a process represents air currents (or any other process related to the phenomenon of interest) well? Sources of definitive information include expert opinion, theoretical results or experimentation. Given an understanding of this information acquired from one of these sources, the designer can then set up the simulation (often by setting parameters in a certain way) to simulate the conditions that produced the information. By studying differences between the output of the simulation and the acquired information the designer can gain insight into how well the simulation is representing the phenomenon of interest. This is an example of using outside information to validate a simulation, an important process associated with insight gathering by the designer.

A designer can spend a great deal of time gathering insight about a phenomenon of interest and then designing and adapting a simulation to properly reflect that phenomenon. In the lifetime of a typical simulation, a significant portion of its existence will probably be spent in an insight gathering role for its designer. Up to this point we have presented the role of modeling and simulation from the perspective of users and designers. In the next section we switch the perspective to simulations, to explore a typical simulation lifetime, moving between the roles of problem solving for its users and insight gathering for its designer.

A SIMULATION'S LIFETIME

Simulations are used as substitutes for phenomena of interest, both real and imagined. They are born when an individual decides it would be useful to possess one. They are used, as we have seen, in both a problem-solving role for users, and an insight-gathering role for designers. Do they play each role once? Twice? Repeatedly? The answer is that different simulations can alternate between the roles different numbers of times, from once to many times. In this section we identify the factors that determine the nature of the lifetime of a simulation.

When the decision is made to create a simulation, a designer needs to reach a point where the phenomenon to be simulated is understood well enough that a meaningful simulation can be created. "Meaningful" can range from "complete, with no designer plans to modify it significantly ever again" to "crude but good enough to start providing insight." The complete case reflects the fact that the designer believes it is possible to develop a valid simulation from the outset. This may or may not happen. If it does, then the simulation will pass from birth to insight gathering to problem solving with no repeats. This sort of simulation development does happen on occasion, but more often the simulation goes through multiple cycles of insight gathering and problem solving phases. Sometimes simulations meant to be developed completely the first time are not realized, perhaps because the designer discovers his or her understanding of the phenomenon of interest is not good enough to finish the simulation design task. In that case the simulation will likely be scrapped or it will pass through an extended role of insight gathering for its designer, or it will be developed through multiple pairs of insight-gathering and problem-solving roles.

For a simulation used primarily in a problem-solving role, the events most likely to trigger its return to a designer insight gathering role include 1) designer realization that s/he now has a better understanding of the phenomenon of interest, 2) user or designer realization that the simulation is not valid for some aspects of the modeled phenomenon, 3) user or designer desire to incorporate some new or remove some existing modeled phenomena in the simulation, and 4) user or designer desire to use the simulation in a context different than any in which it was originally designed to operate validly. In all cases the simulation is returned to an insight-gathering role so that it may be adapted to meet the new requirements associated with the triggering condition.

For a simulation used primarily in an insight-gathering role, the event that will most likely trigger its transition to a problem-solving role is when the simulation is finally deemed valid for its intended uses. A simulation transitioning away from a designer insight-gathering role is generally calibrated using known or hypothesized data, before it is placed into a problem-solving role.

Simulations can easily have lifetimes that span decades. Many of these simulations exist. Some have not been changed for years, having entered a stable problem solving role. Others cycle between problem-solving and insight-gathering roles on a regular basis, for all of the triggering conditions described above. In many cases individuals or small groups of researchers own decades old simulations and regard them as a part of their ongoing research. As new insight is gained, by way of simulation, theory or experimentation, a simulation will be modified, revalidated (employing designer insight gathering) for intended uses, possibly calibrated, and then brought back into the ongoing research project for further problem solving.

Some simulations are developed for one-time use, but they tend to be more rare. In many cases a simulation intended for one time use eventually gets used many times, and enters the simulation lifecycle of transitions between problem solving and designer insight gathering, perhaps eventually becoming stable in a problem-solving role.

CONCLUSION

We have seen that simulations are used in the roles of problem solving and insight gathering. Which role a simulation occupies is determined mainly by who is using it (user or designer), how much epistemic uncertainty exists with respect to the phenomenon of interest, and how much uncertainty exists with respect to how well the simulation represents the phenomenon of interest. The more we can reduce the epistemic uncertainty associated with a simulated phenomenon, the more likely the simulation can be used for problem solving without requiring phases of designer insight gathering to improve the model. We can often determine the current role of a simulation by studying the questions it is being used to answer. Finally, in its life-time, a typical simulation transitions between problem solving and insight gathering roles as new information about the phenomenon of interest is learned, or users wish to have the simulation meet new requirements.

KEY TERMS

prediction
model calibration

aleatory uncertainty
epistemic uncertainty

gaining insight

FURTHER READING

CIPRA BA. Revealing uncertainties in computer models. *Science* 2000;287(5455):960–961.

DAREMA F. Dynamic data driven applications systems: A new paradigm for application simulations and measurements. In Bubak M, van Albada GD, Sloot PMA, and Dongarra JJ, eds. *Computational Science—ICCS 2004: 4th International Conference*, volume 3038, Heidelberg, Germany: Springer-Verlag; 2004.

DAVIS PK. New paradigms and new challenges. In Kuhl ME, Steiger NM, Armstrong FB, and Joines JA, eds. *Proceedings of the 2005 Winter Simulation Conference*, pp. 1067–1076, Piscataway, NJ: Institute of Electrical and Electronics Engineers, Inc.; 2005.

DAVIS PK, and BIGELOW JH. Motivated metamodels: synthesis of cause-effect reasoning and statistical metamodeling. *Technical Report MR-1570-AF*. RAND Corporation; 2003.

ELDERD BD, DUKIC VM, and DWYER G. Uncertainty in predictions of disease spread and public health responses to bioterrorism and emerging diseases. *Proceedings of the National Academy of Science* 2006;103(42)15693–15697.

EPSTEIN JM and AXTELL R. *Growing Artificial Societies: Social Science from the Bottom Up*. Washington, DC: Brookings Instititution Press; 1996.

FERGUSON NM, KEELING MJ, EDMUNDS WJ, GANI R, GREENFELL BT, and ANDERSON RM. Planning for smallpox outbreaks, *Nature* 2003;425:681–685.

FUJIMOTO R, LUNCEFORD WH, PAGE EH, UHRMACHER AM, eds. Grand challenges for modeling and simulation. Number 350. Saarbrücken, Germany: Schloss Dagstuhl; 2002.

HELTON JC, JOHNSON JD, OBERKAMPF WL, and STORLIE CB. A sampling-based computational strategy for the representation of epistemic uncertainty in model predictions with evidence theory. Alburquerque, NM: Sandia National Laboratories; 2006.

LAW AM and KELTON WD. *Simulation Modeling and Analysis*. Boston: McGraw Hill; 2006.

National Science Foundation. Simulation-based engineering science: Revolutionizing engineering science through simulation. *Report of the NSF Blue Ribbon Panel on Simulation-Based Engineering Science*, February, 2006.

OBERKAMPF WL, HELTON JC, JOSLYN CA, WOJTKIEWICZ SF, and FERSON S. Challenge problems: uncertainty in system response given uncertain parameters. *Reliability Engineering & System Safety* 2004;85:11–19.

OCRWM. Yucca Mountain Science and Engineering Report REV 1. DOE/RW-0539-1. Las Vegas, NV: U.S. Department of Energy, Office of Civilian Radioactive Waste Management; 2002.

PARNAS DL and WURGES H. Response to undesired events in software systems. In *Proceedings of the Second International Conference on Software Engineering*, pp. 437–447, Oct 1976.

SPIEGEL M, REYNOLDS PF, and BROGAN DC. A case study of model context for simulation composability and reusability. In Kuhl ME, Steiger NM, Armstrong FB, and Joines JA, eds. *Proceedings of the 2005 Winter Simulation Conference.* Piscataway, NJ: Institute of Electrical and Electronics Engineers, Inc.; 2005.

GNEITING T and RAFTERY AE. Weather forecasting with ensemble methods. *Science* 2005;310:248–249.

ZEIGLER BP, PRAEHOFER H, and KIM TG. *Theory of Modeling and Simulation*, 2nd Edition. Burlington, MA: Academic Press; 2000.

Part Two

Theoretical Underpinnings

Chapter 3

Simulation: Models That Vary over Time

John A. Sokolowski

INTRODUCTION

Recall that the concept of a model is, by definition, a static representation of that world. Simulation adds a temporal aspect to a model by depicting how the system being modeled changes over time. So simulation can be defined as a *time-varying* representation of a model. One may also think of simulation as a process by which you perform experiments on a model to see how a real system would behave if it had the same experiments performed on it. For example, aircraft designers build scale models of airplanes to test the aerodynamic characteristics of their designs. By manipulating the wing design or the shape of the fuselage on the model the designer can ascertain how his or her changes will affect the air flow over these surfaces. If the model is a true enough representation of the actual aircraft then the designer will have evidence of how the real system would behave had these changes been made to it. You can see the advantage of simulation in that you may conduct multiple experiments on a model under varying conditions without having to incur the expense of building and modifying the actual system. It is much easier to change the characteristics of a model in a simulation than it is trying to refabricate a full-scale aircraft.

This chapter will look at two types of simulation, *discrete event simulation* and *continuous simulation*. Each type has a distinct way of adding a time aspect to a model. Discrete event simulation relies on the occurrence of specific events to advance a model from one state to another over time, while continuous simulation represents model behavior over a continuum of time independent of events occurring in the system. As an illustration of the difference between these two types of simulation consider a model of a traffic light and a car. The traffic light can exist in three

Principles of Modeling and Simulation: A Multidisciplinary Approach, Edited by John A. Sokolowski and Catherine M. Banks.

states: red, yellow, or green. It usually changes state when an internal timer counts down a set amount of time and triggers an event to change the light. A car, on the other hand, cannot instantaneously change from one speed to another. It must undergo an acceleration, which is defined as a change in speed over time. This change in speed is a continuous time event rather than a single instantaneous occurrence. The following discussion presents the methodology for implementing these two types of simulation.

DISCRETE EVENT SIMULATION

We will begin by studying the concepts that make up discrete event simulation. We formally define **discrete event simulation** as the variation in a model caused by a chronological sequence of events acting on it. **Events** are instantaneous occurrences that may cause variations or changes in the state of a system. The **state of a system** is defined as one or more variables that completely describe a system at any given moment in time. These variables are called **state variables**.

Take, for example, the traffic light from above. Its state can be represented by which light is activated at any given time. So a state variable called *light color* is chosen to represent its state. The value of that variable at any given point in time completely describes the state of the traffic light. Events for the traffic light system consist of *switch to red, switch to yellow*, and *switch to green*. These events occur in a predetermined sequence and may be triggered by the passage of a certain amount of time. Or they may be triggered by the event of the presence of a vehicle over some roadway sensor or a video system recognizing when a vehicle enters its field of view.

Another component of a discrete event simulation system is a clock. The clock keeps track of **simulation time** and may be used to trigger events in the system such as when to switch colors from red to green. Simulation time may or may not be the same as *wall clock* or real time. Wall clock time refers to how humans perceive the passage of time. A simulation may run at wall clock time to provide a realistic representation as perceived by a human. It may also run considerably faster than wall clock time to allow for analysis of a system that would normally take hours, weeks, or months to evolve in real time. The ability to study the behavior of a system in a compressed amount of time is one advantage of using simulation. The traffic light is a very simple illustration of a discrete event simulation. Let us look at a more complex example to introduce other discrete event simulation concepts.

Almost everyone has stood in line in a store or in a bank waiting to be served. The length of time we must wait is influenced by the number of people ahead of us, the number of cashiers or tellers servicing customers, and the length of time it takes for each customer to be serviced. This type of system is known as a **queuing system** and it is representative of many types of systems found in our daily lives. Typical events that you would find in a queuing system are the arrival of a customer, the servicing of a customer, and the departure of a customer. Figure 3-1 shows some

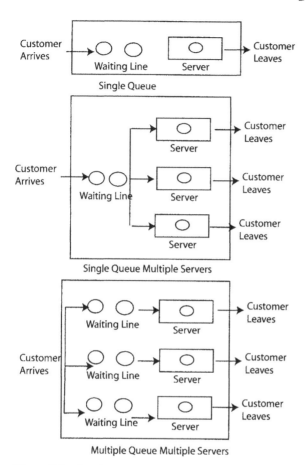

Figure 3-1   Queuing system configurations

common queuing system configurations. The top system is a single queue/single server system. The middle system represents a single queue/multiple server system. The bottom system depicts a multiple queue/multiple server system. Queuing systems lend themselves very well to discrete event simulation. This chapter describes discrete event simulation in general. Chapter 4 will provide a more in-depth look at queuing systems since they represent a large segment of where discrete event simulation may be used.

Customer arrival is usually a **random event**. Random event refers to occurring without a recognizable pattern. Random events can be represented by statistical distributions that allow one to simulate these seemingly random occurrences. Customer service time may also be random since it may depend on the amount of grocery items the customer has or the complexity of the banking transaction that must occur. In a discrete event simulation these random events are controlled by an **event list**, which is a list that contains the next time an event will occur. The time

Table 3-1   Queuing System Partial Event List

Event Time	Event
00:17	Arrival of customer 1
00:22	Arrival of customer 2
00:24	Servicing of customer 1
00:26	Arrival of customer 3
00:27	Departure of customer 1
00:28	Servicing of customer 2

for each event is randomly chosen by an **event generator** and added to the event list. The event list is ordered in increasing time. When the simulation clock matches the time of the next event in the event list, that event is executed and the state of the system will instantaneously change to a new state. That event is then removed from the event list and the next event waits to be executed. Table 3-1 is an example of a partial event list for a queuing system.

You can see from the event list of Table 3-1 that customer 1 arrives at time 00:17, customer 2 arrives at time 00:22, and thus a queue of customers begins to build waiting for service. At time 00:24 customer 1 is serviced. The simulation then progresses as each event is processed. The state of the system at any given time is represented by the customers waiting in the service queue, the customers receiving service, and the customers that have departed. Thus there would be a state variable for each one of these items. The event list is not considered a state variable since it does not represent the current system condition. Instead it represents what will happen in the future and controls how the simulation changes state with the passage of time.

Representing Discrete Event Simulations in a Digital Computer

Computers lend themselves very well to discrete event simulation. Since digital computers are inherently discrete machines that change state based on some discrete timing sequence they can readily handle the execution of a simulation in this discrete manner. The computer's clock is used to control the simulation clock and thus the processing of events in the event list. We will take advantage of a computer's ability to simulate a discrete event system by employing a spreadsheet program to construct a discrete event simulation and capture the state of the system as it evolves over time.

Using a Spreadsheet for Simulation

Spreadsheet programs are widely available and serve as a handy tool for constructing many types of simulations. In this section a spreadsheet will be used to build a discrete event simulation. Cells of a spreadsheet can accept many types of data like text, numeric, and formula entries. Text data are such things as column headings. Numeric data are specific numeric values. Formulas take advantage of numeric data, operators such as multiplication, and functions to compute new values. Columns of a spreadsheet can be thought of as lists that can hold a sequence of events, thus they can represent the event list. Rows in a spreadsheet can represent the state of the system with each row containing specific values for each state variable (column headings) at a given point in time. A spreadsheet will be used to implement a discrete event queuing system.

Example 1: Simulation of a Single-Server Queuing System

A local bank has one teller who provides banking services to customers arriving throughout the day. When the customers arrive they must wait in a single-file line to be serviced by the teller. The customers arrive randomly throughout the day and the time it takes for each customer to be serviced is also random. Figure 3-2 depicts the single server (teller) queuing system.

Note that there are three main processes in the system, the *arrival process*, the *service process*, and the *departure process*. The arrival process is often described by a random variable with a Poisson statistical distribution.[1] This distribution is mathematically defined as:

$$P(x) = \frac{\lambda^x e^{-x}}{x!}, \; for \; x = 0, 1, 2, \ldots \tag{3.1}$$

where λ is the arrival rate (e.g., customers arrive at a rate of λ per hour) and x represents the number of customers. Thus, $P(0)$ is the probability that zero customers will arrive in any given hour. $P(1)$ is the probability of one customer arriving in any given hour and so on. If one knows by observation that the average customer arrival rate is five customers per hour ($\lambda = 5$) then the probability distribution would look like Figure 3-3.

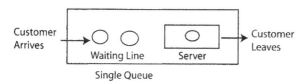

Customer Arrives → Waiting Line Server → Customer Leaves

Single Queue

Figure 3-2 Single-server queuing system

---

[1] Siméon-Denis Poisson (1781–1840) determined a probability theory of distribution. His research centered on assessing certain random variables N that count a number of discrete occurrences that take place during a time-interval of given length. Poisson published his findings in his essay, *Research on the Probability of Judgments in Criminal and Civil Matters*. The *Poisson distribution* is a discrete probability distribution that expresses the probability of a number of events occurring in a fixed period of time. This is possible only if these events occur with a known average rate and if these events are independent of the time since the last event.

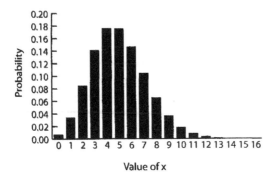

Value of x

Figure 3-3 Poisson Probability Distribution for customer arrival rate of five per hour

	B5			f_x	=($B$2^A5*EXP(-$B$2))/FACT(A5)					
	A	B	C	D	E	F	G	H	I	J
1	Average Arrival Rate:									
2	λ=	5	per hour							
3										
4	x	P(x arrivals)								
5	0	0.0067								
6	1	0.0337								
7	2	0.0842								
8	3	0.1404								
9	4	0.1755								
10	5	0.1755								
11	6	0.1462								
12	7	0.1044								
13	8	0.0653								
14	9	0.0363								
15	10	0.0181								
16	11	0.0082								
17	12	0.0034								
18	13	0.0013								
19	14	0.0005								
20	15	0.0002								
21	16	0.0000								
22	Sum	1.000								
23										
24										

Poisson

Figure 3-4 Spreadsheet showing the Poisson Probability calculation

A spreadsheet can be used to generate the probabilities for any given average arrival rate by embedding equation 3.1 into the spreadsheet cells and automatically performing the calculations for us. Figure 3-4 shows the setup of the spreadsheet for this calculation.

Cell B2 holds the average arrival rate value. In this example the value is 5. Cell B5 contains equation 3.1 using the mathematical operators and functions built into the spreadsheet program (=($B$2^A5*EXP(-$B$2))/FACT(A5)). Note that column B from B5 to B21 contains the x values for which to evaluate the Poisson probability.

So the simulation now knows the probability that x number of customers will arrive in any given hour with an average expected arrival of 5. As Figure 3-4 illustrates, since the average arrival is 5 then the highest probability of arrival is also 5. But that probability is only about 0.175 since it is possible that more or less than five will arrive each hour. That is why customer arrival is described as a random variable.

The spreadsheet easily allows for the calculation of the Poisson probability distribution for other average arrivals per hour. Cell B2 can be changed to whatever that average arrival is and the spreadsheet will automatically recompute the probabilities.

Interarrival Time is the time between customer arrivals. This time is important because it will indicate the random time at which the next customer will arrive at the queue. This number will need to be computed for the simulation to account for the arrival of each customer. The Interarrival Time is exponentially distributed and follows the probability function of equation 3.2.

$$P_0(t) = e^{-\lambda t} \tag{3.2}$$

$P_0(t)$ is the random probability and t is the Interarrival Time. Equation 3.2 can be solved for the Interarrival Time. Its solution is as follows.

$$t = \frac{-1}{\lambda} \ln(P_0(t)) \tag{3.3}$$

The service process is described by a different probability distribution, the *Exponential Random Variable*. This random variable is given by the following equation.

$$P(t_1 \leq T \leq t_2) = e^{-\mu t_1} - e^{\mu t_2} \text{ for } t_1 < t_2 \tag{3.4}$$

where μ is the service rate. For example customers can be serviced at a rate of $\mu = 7$ per hour. This implies that the average service time is $1/\mu$ or about 8.5 minutes. $P(t_1 \leq T \leq t_2)$ in equation 3.4 represents the service time interval probability or what is the probability that a customer's service time will be between t_1 and t_2. The spreadsheet can again be used to produce a probability distribution for this example. The setup for this spreadsheet is shown in Figure 3-5.

Note in Figure 3-5 that increments of 0.05 hours were chosen as service intervals. This increment equates to service intervals of every 3 minutes. So from Figure 3-5 the probability of the service time being between 0 and 3 minutes is 0.295, between 3 and 6 minutes is 0.208 and so on. These probabilities are cumulative so the probability of the service time being between 0 and 6 minutes would be 0.295 + 0.208 or 0.503 for the known average service rate of 7 per hour. Again the service rate may be changed to match the behavior of a specific system and the spreadsheet will automatically recalculate the probability distribution.

The probability of a specific service time is computed in a similar manner to the interarrival time and follows the same function as equation 3.3. Equation 3.3 will be used to compute each service time for the server (teller) simulation. The departure process could also have some probability of delay associated with it. For

	D5			f_x	=EXP(-$C$2*A5)-EXP(-$C$2*C5)				⮝

	A	B	C	D	E	F	G	H	I
1	Average Service Rate:								
2	μ =	7		per hour					
3									
4	Service Time			Probability					
5	0.00	to	0.05	0.295					
6	0.05	to	0.10	0.208					
7	0.10	to	0.15	0.147					
8	0.15	to	0.20	0.103					
9	0.20	to	0.25	0.073					
10	0.25	to	0.30	0.051					
11	0.30	to	0.35	0.036					
12	0.35	to	0.40	0.025					
13	0.40	to	0.45	0.018					
14	0.45	to	0.50	0.013					
15	0.50	to	0.55	0.009					
16	0.55	to	0.60	0.006					
17	0.60	to	0.65	0.004					
18	0.65	to	0.70	0.003					
19	0.70	to	0.75	0.002					
20	0.75	to	0.80	0.002					
21	0.80	to	0.85	0.001					
22	0.85	to	0.90	0.001					
23	0.90	to	0.95	0.001					
24			Sum	1.00					
25									
26									

⏮ ◀ ▶ ⏭	Exponential	⏏			

Figure 3-5   Spreadsheet showing exponential distribution calculation

simplicity of this problem it is assumed that once a customer is serviced he or she immediately departs the system.

There is one other aspect that must be considered before the full discrete event simulation can be set up for the bank teller system. Recall that the arrival rate and service time are random variables. This implies that the simulation must be capable of generating random numbers to select specific values for these random variables. Random number generation in a discrete event simulation is a very important concept because it forms the basis for creating the event list which associates events with specific times just as was shown in Table 3-1. Many algorithms exist to generate what are known as *pseudo random numbers*. The term *pseudo* is used because these algorithms are not truly random. The better algorithms will generate millions of numbers before a pattern is created. In most cases this is sufficient to give the appearance of true randomness. Spreadsheets have built in random number generation capability. For small discrete event simulations this capability is adequate. See McCullough and Wilson [1] for an analysis of random number generation in a typical spreadsheet.

To set up the spreadsheet simulation for the bank teller system the following spreadsheet columns will be used: *Customer Number, Interarrival Time, Arrival*

Time, Number in Queue, Service Start Time, Service Time, and *Completion Time*. *Mean (average) Arrival Time* and *Mean (average) Service Rate* will also be set. Given these variables the spreadsheet setup should look like Figure 3-6.

The next step will be to specify the mean arrival rate, the mean service rate, and the number of customers to be simulated. The mean arrival and mean service rates will be the ones illustrated above. The setup should now look like Figure 3-7.

The arrival of each customer and his or her service time is the data necessary to build the event list. So Figure 3-8 will now be filled in to provide those times.

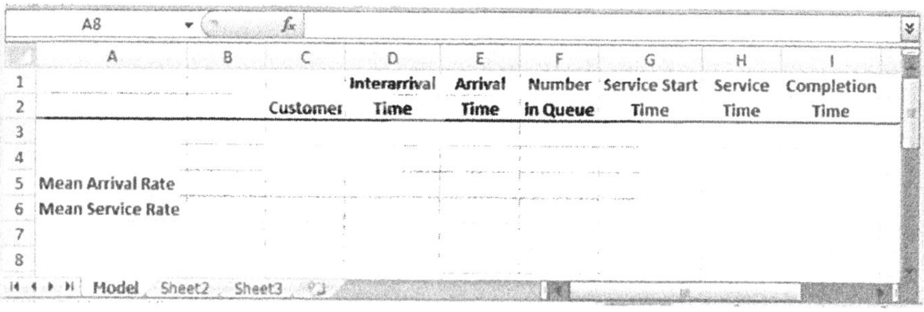

Figure 3-6 Queuing system spreadsheet setup

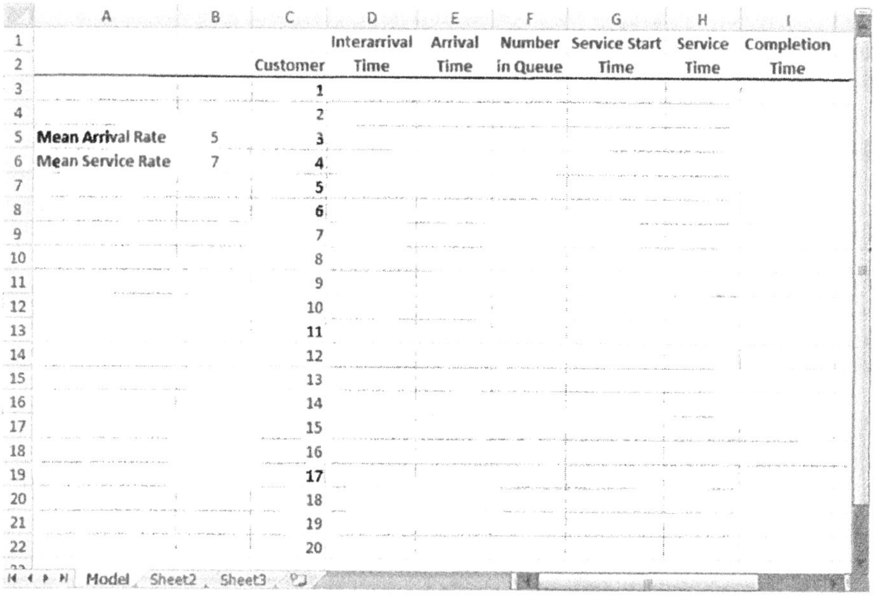

Figure 3-7 Queuing system with Mean Arrival Rate, Mean Service Rate, and Customers

	D4		▼	f_x	=-(1/$B$5)*LN(RAND())				
	A	B	C	D	E	F	G	H	I
1				Interarrival	Arrival	Number	Service Start	Service	Completion
2			Customer	Time	Time	in Queue	Time	Time	Time
3									0
4			1	0.08226079				0.01675	
5	Mean Arrival Rate	5	2	0.17232941				0.325443	
6	Mean Service Rate	7	3	0.29510383				0.060776	
7			4	0.24580561				0.10761	
8			5	0.35051459				0.10012	
9			6	0.24564337				0.012546	
10			7	0.17713883				0.011608	
11			8	0.39191625				0.226435	
12			9	0.06566114				0.286761	
13			10	0.10298295				0.047811	
14			11	0.03716799				0.132653	
15			12	0.07029501				0.024532	
16			13	0.28347553				0.119972	
17			14	0.27778723				0.137517	
18			15	1.20450807				0.081098	
19			16	0.05951389				0.098972	
20			17	0.7085237				0.019817	
21			18	0.00157035				0.001505	
22			19	0.14082754				0.025756	
23			20	0.12325655				0.040887	
24									

◄ ◄ ► ► Model   Sheet2   Sheet3

Figure 3-8   Calculation of Interarrival Time

Note the calculation of the Interarrival Time of customer #1 is based on equation 3.3 as can be seen in the formula box of Figure 3-8. Cell D4 indicates that Interarrival Time. This arrival time assumes the simulation started at time zero. Service times were generated in a similar manner using equation 3.3. This provides the simulation with the calculations for all the random variables needed for the first twenty customers. The remaining column values are calculated as follows. *Arrival time* is the sum of the previous arrival time plus the Interarrival Time for the next customer. *Number in Queue* is calculated using the following spreadsheet formula illustrated for cell F5: =C5-MATCH(E5,$I$3:I4,1). *Service Start Time* is the maximum value (MAXA) between the current arrival time and the previous completion time. Finally, *Completion Time* is the sum of the service time and service start time. Following the above process produces the complete spreadsheet of Figure 3-9.

Using Figure 3-9 the event list for the bank server (teller) simulation can now be constructed. Table 3-2 provides a partial list of events for the first eleven customers.

Note that the first seven customers arrived when the teller was not servicing another customer so their service began as soon as they arrived. However, customer #8 had to wait because customer #7 was still being serviced. This is also true for customers 9, 10, and 11. Figure 3-9 shows the number in the queue building because of the length of time it is taking for customer #8 to be serviced.

	A	B	C	D	E	F	G	H	I
				Interarrival	Arrival	Number	Service Start	Service	Completion
			Customer	Time	Time	in Queue	Time	Time	Time
3									0
4			1	0.4596	0.4596	0.0000	0.4596	0.0823	0.5418
5	Mean Arrival Rate	5	2	0.2233	0.6829	0.0000	0.6829	0.1267	0.8095
6	Mean Service Rate	7	3	0.1507	0.8335	0.0000	0.8335	0.0038	0.8373
7			4	0.0687	0.9022	0.0000	0.9022	0.1082	1.0105
8			5	0.2364	1.1387	0.0000	1.1387	0.0313	1.1699
9			6	0.0885	1.2271	0.0000	1.2271	0.0463	1.2734
10			7	0.2037	1.4308	0.0000	1.4308	0.1188	1.5496
11			8	0.0677	1.4986	1.0000	1.5496	0.4709	2.0205
12			9	0.0649	1.5634	1.0000	2.0205	0.0059	2.0264
13			10	0.2216	1.7850	2.0000	2.0264	0.0336	2.0600
14			11	0.2031	1.9881	3.0000	2.0600	0.0101	2.0701
15			12	0.0538	2.0419	2.0000	2.0701	0.1602	2.2303
16			13	0.2314	2.2732	0.0000	2.2732	0.1425	2.4157
17			14	0.4311	2.7043	0.0000	2.7043	0.1251	2.8294
18			15	0.6192	3.3235	0.0000	3.3235	0.0279	3.3514
19			16	0.1754	3.4989	0.0000	3.4989	0.1259	3.6248
20			17	0.0066	3.5055	1.0000	3.6248	0.0607	3.6855
21			18	0.0624	3.5680	2.0000	3.6855	0.0408	3.7262
22			19	0.5204	4.0883	0.0000	4.0883	0.2162	4.3045
23			20	0.0196	4.1079	1.0000	4.3045	0.2813	4.5858

Figure 3-9   Customer event data

Using a spreadsheet simulation like this example provides an easy means to simulate an entire day, week, or month of bank operation to determine if the number of tellers is adequate to keep the queue size small and the customer wait times at an acceptable level. Many types of statistics can be computed for this analysis. The simulation may also be updated at any point in time if the average arrival rate or average service time changes to show the impact on the system.

This type of discrete event simulation can be expanded to analyze increasingly complex systems. For example, the system may have multiple tellers, each one with his or her own average service time. Customer arrival rate may change depending on the time of day or day of the week. Even with these complexities the spreadsheet model may still be employed to simulate the behavior of the system and to gain valuable information on its operation and efficiency.

Thus far you have been introduced to discrete event simulation and its concepts. In the server (teller) example many state changes occurred as customers arrive, were serviced, and departed the system. These changes are captured in the event list, which provides a temporal order for the system to progress from one state to another. Many other types of systems may be simulated in this manner. Students are encouraged to use the techniques presented here to construct their own simulations of systems that interest them. Let's now look at continuous simulation.

Table 3-2   Bank Teller Simulation Event List

Time	Event
0.4596	Customer 1 Arrives
0.4596	Customer 1 Begins Service
0.5418	Customer 1 Departs
0.6829	Customer 2 Arrives
0.6829	Customer 2 Begins Service
0.8095	Customer 2 Departs
0.8335	Customer 3 Arrives
0.8335	Customer 3 Begins Service
0.8373	Customer 3 Departs
0.9022	Customer 4 Arrives
0.9022	Customer 4 Begins Service
1.0105	Customer 4 Departs
1.1387	Customer 5 Arrives
1.1387	Customer 5 Begins Service
1.1699	Customer 5 Departs
1.2271	Customer 6 Arrives
1.2271	Customer 6 Begins Service
1.2734	Customer 6 Departs
1.4308	Customer 7 Arrives
1.4308	Customer 7 Begins Service
1.4986	Customer 8 Arrives
1.5496	Customer 7 Departs
1.5496	Customer 8 Begins Service
1.5634	Customer 9 Arrives
1.7850	Customer 10 Arrives
1.9881	Customer 11 Arrives

CONTINUOUS SIMULATION

As was shown in the previous section discrete event simulation is based on a chrono-logical sequence of events that caused the model to change state each time an event occurred. **Continuous simulation,** on the other hand, is defined as a system described by state variables that change continuously with respect to time. *State variables* are parameters of the system that describe the system's behavior. Consider the motion of an object such as a ball dropped from a tower several hundred feet in the air. One state variable that describes the motion of this ball is velocity. Before the ball is released it has zero velocity. Once released, the ball begins to accelerate due to the force of gravity. The ball's velocity increases in a smooth fashion with no discern-able pause. Figure 3-10 is a graph of this ball's velocity over time assuming only the force of gravity acting on it.

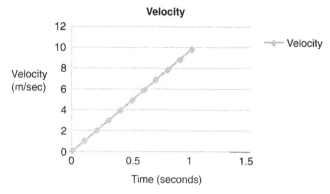

Figure 3-10   Velocity of dropped ball

Note the continuous nature of the curve. This curve was produced by a simulation that calculates the ball's velocity in a continuous manner so that it very closely represents the actual physical system behavior. If we dropped a ball and had a way of measuring and plotting its velocity as it fell, the two curves would be nearly identical assuming the model accounted for all the forces acting on the ball as it fell. The accuracy of that assumption is clearly dependent on how well the model represents that actual system. (The concept of verification and validation, or the measure of how closely the model represents reality, will be discussed in Chapter 6.)

Continuous simulation is often used to simulate physical systems such as this ball or systems that involve mechanical, electrical, thermal, or hydraulic components. However, it can be used to represent any continuous time-varying system. For example, the spread of a disease such as the flu can be depicted in this manner. As the disease is introduced into a population the number of infected people starts to increase. In large populations the graph representing the number of infected people appears as a continuous function and can be represented mathematically as such. As the population begins to recover and becomes immune to the disease the number of infected people decreases. This pattern continues until no susceptible population remains and the disease dies out. Figure 3-11 shows an illustration of a typical disease spread in a regional population. Note the continuous nature of the curves.

Representing Continuous Simulations in a Digital Computer

From the examples above we saw continuous functions that described the response of the system at any point in time. One can develop mathematical models to represent these continuous functions. These models consist of one or more equations that include the state variables of the system. *Differential equations* are the most precise

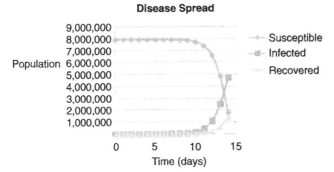

Figure 3-11   Disease spread illustration

form of these equations. However, they require an understanding of calculus. Instead, we will use an algebraic substitute for differential equations known as **difference equations**. Difference equations will allow the use of algebra and a spreadsheet to construct a simulation of a continuous system.

Using the ball example the difference equation that defines its velocity is:

$$v(t + \Delta t) = v(t) + \alpha(t) \Delta t \tag{3.5}$$

where $v(t + \Delta t)$ is the velocity of the ball at time $t + \Delta t$, $v(t)$ is the velocity of the ball at some time t, and $a(t)$ is the acceleration of the ball at time t. What this equation shows is that the future velocity of the ball depends on its current velocity and how much acceleration the ball possesses at a given point in time.

Difference equations are very convenient for developing a simulation on a digital computer. By its very nature a digital computer operates on discrete time steps. Digital computers contain internal clocks that regulate the execution of instructions at a set interval of time. Since the difference equations contain discrete time steps in the form of Δt we can readily compute new values for the state variables based on these discrete advances in time. As the time steps become smaller and smaller, a continuous curve representation of the state variable is approached. This is the concept that will be employed to produce a continuous simulation representation of the system.

One other aspect of the method for generating a continuous simulation on a digital computer must be considered. Suppose the value for acceleration (a) in equation (3.5) is constant. Then we can compute the value of any future velocity of the ball by taking its present velocity and adding the product of the ball's constant acceleration and Δt. From an equation standpoint we would have:

$$v(t + \Delta t) = v(t) + \text{area under the } \alpha \text{ vs. } t \text{ curve} \tag{3.6}$$

or

$$v(t + \Delta t) = v(t) + \alpha(t_1)(t_2 - t_1) \tag{3.7}$$

Figure 3-12 depicts this calculation graphically. Note the hashed area represents the change in velocity over this time interval.

Now suppose that acceleration is not constant. Instead it is defined by the following curve in Figure 3-13.

We can still approximate the change in acceleration by the hashed rectangle representing the area under the a vs. t curve. But this rectangular area has some error associated with it given by the space above the rectangle and below the acceleration curve. So we will not be able to compute the exact value for accel-

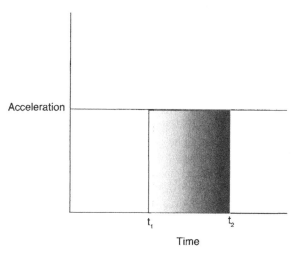

Figure 3-12 Constant acceleration graph

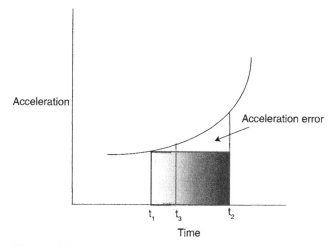

Figure 3-13 Acceleration error characteristic

eration. Note, however, as we make the time interval Δt smaller, this error becomes smaller. If we choose an appropriately small value for Δt the amount of error in the approximation can be controlled. This process is known as *Euler's Method* used for estimating area under a curve.[2] It will allow for determining the *instantaneous acceleration* or acceleration at any given point in time on the curve. This method will be employed to develop continuous simulations on a digital computer.

Using a spreadsheet a continuous simulation of the falling ball will be constructed based on equation 3.5 to plot its velocity over time for a given time interval. To do this we will describe how to set up a spreadsheet to support building this continuous simulation.

Example 2: Simulating a Falling Ball

For the spreadsheet model we will need a column representing *Time*, a column representing the *Future Velocity* $v(t + \Delta t)$, a column for the *Current Velocity* $v(t)$, and a column for the *Change in Velocity* $\alpha(t_1)(t_2 - t_1)$. We may also want some cells to hold constant values in the formula so that we don't have to retype them each time. For this example we will assume we have a constant acceleration of the ball due to gravity and we will have a cell to hold that constant value. Another constant value that should be considered is that of the time increment Δt. By choosing different values for the time increment, we can observe its effect on the accuracy of the simulation. Remember, there may be an area under a curve that we are not including and that area is controlled by the time increment selection. The set up of the spreadsheet should look like that of Figure 3-14.

Now let's look at the state of the ball at time zero just before it was dropped. The ball is not moving so its velocity is zero. The force of gravity is pulling on it. That force is 9.8 meters per second per second or 9.8 *meters/sec*^{2}. Its future velocity will be calculated from equation 3.5. We will use a time increment of 0.1 seconds. So the spreadsheet should now look like Figure 3-15.

Note that in cell D2 we compute the value for the change in velocity by multiplying the acceleration by the time increment $\alpha(t_1)(\Delta t)$. The spreadsheet allows for

	A	B	C	D	E	F	G
1	Time	Future Velocity	Current Velocity	Change in velocity (acceleration)		Time Increment	Acceleration
2							
3							
4							

Sheet1 Sheet2 Sheet3

Figure 3-14 Spreadsheet simulation setup

2 *Euler's Method* is used in mathematics and computational science as the first order numerical procedure for solving ordinary differential equations with a given initial value.

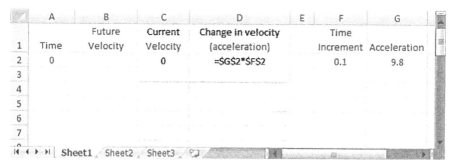

Figure 3-15   Time zero spreadsheet values

	A	B	C	D	E	F	G
		Future	Current	Change in velocity		Time	
1	Time	Velocity	Velocity	(acceleration)		Increment	Acceleration
2	0		0	=$G$2*$F$2		0.1	9.8
3							
4							
5							
6							
7							

Figure 3-16   Spreadsheet computation for future velocity

	A	B	C	D	E	F	G
		Future	Current	Change in velocity		Time	
1	Time	Velocity	Velocity	(acceleration)		Increment	Acceleration
2	0	=C2+D2	0	0.98		0.1	9.8
3							
4							
5							

Figure 3-17   Spreadsheet simulation after one-time step

	A	B	C	D	E	F	G
		Future	Current	Change in velocity		Time	
1	Time	Velocity	Velocity	(acceleration)		Increment	Acceleration
2	0	0.98	0	0.98		0.1	9.8
3	0.1	1.96	0.98	0.98			
4							
5							

embedding the formula for that calculation in the cell (=$G$2*$F$2). Finally, we compute the future velocity $v(t)$ by embedding equation 3.5 in the future velocity cell. This step is shown in Figure 3-16.

We will now set up the spreadsheet for the velocity computation for the next time increment, which will be time 0.1 seconds. Note that at time 0.1 seconds the current velocity becomes the future velocity from time zero. So the spreadsheet simulation should now resemble Figure 3-17.

We can now continue the spreadsheet simulation for any length of time we desire to see how the ball's velocity will change over time. We would expect the ball's velocity to steadily increase over time because of the constant value for

	A	B	C	D	E	F	G
		Future	Current	Change in velocity		Time	
1	Time	Velocity	Velocity	(acceleration)		Increment	Acceleration
2	0	0.98	0	0.98		0.1	9.8
3	0.1	1.96	0.98	0.98			
4	0.2	2.94	1.96	0.98			
5	0.3	3.92	2.94	0.98			
6	0.4	4.9	3.92	0.98			
7	0.5	5.88	4.9	0.98			
8	0.6	6.86	5.88	0.98			
9	0.7	7.84	6.86	0.98			
10	0.8	8.82	7.84	0.98			
11	0.9	9.8	8.82	0.98			
12	1	10.78	9.8	0.98			

Sheet1 Sheet2 Sheet3

Figure 3-18   Spreadsheet simulation after one second of simulation time

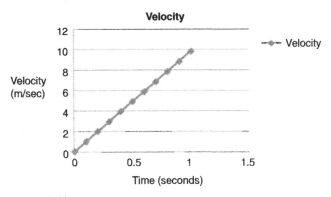

Figure 3-19   Velocity of falling ball

acceleration. Figure 3-18 shows the calculation of velocity for a period of 1 second.

Note that at one second its velocity is 9.8 meters per second. This is something we would expect with the constant acceleration value that was used. We can employ the spreadsheet's built-in graphing capability to produce a graph of the ball's velocity over time (Figure 3-19). The graph clearly shows the continuous nature of the ball's velocity.

We now have a method to construct a continuous simulation using common spreadsheet functionality. Remember, however, that the simulation is only as good as the underlying model. Ensuring that the simulation matches the real-world system to a specified degree of accuracy is part of the *verification and validation*.

Example 3: Simulating a Biological System

Mechanical systems such as the ball are not the only type of system represented by continuous simulation. Recall we mentioned that the progression of a disease through a population can also be represented in this manner. So let's use the spreadsheet method to construct a simulation of a typical disease cycle.

When modeling the spread of a disease we normally consider three groups of people: those susceptible to the disease, those infected by the disease, and those who have recovered from the disease. There are many variations on these groups but these three populations are most commonly used when studying disease spread. We choose the following variables to represent these three groups: S—Susceptible, I—Infected, and R—Recovered.

There are two other terms we must define. *Infection rate* (λ) represents the number of people who become infected per unit time. We must also consider *recovery rate* (δ), or how long it takes for an individual to get over the disease. We are now ready to develop the difference equations to represent this biological system.

$$S(t + \Delta t) = S(t) - \lambda S(t) I(t)(\Delta t) \tag{3.8}$$

$$I(t + \Delta t) = I(t) - (\lambda S(t) I(t) - \delta I(t))(\Delta t) \tag{3.9}$$

$$R(t + \Delta t) = R(t) - \delta I(t)(\Delta t) \tag{3.10}$$

Equation 3.8 indicates that the susceptible population decreases (becomes infected) based on how many people within the uninfected population become infected, how many infected people who are in the population who can pass on the disease, and the rate at which the disease is transferred. The infected population changes based on how many are becoming infected and how many have recovered (equation 3.9). Finally, the recovered population is controlled by the number of infected multiplied by the recovery rate (equation 3.10).

We will use a real-world example for the simulation. In the winter of 1968–1969 a strain of flu known as the Hong Kong flu swept the United States. We will simulate the spread of the disease in New York City with its then population of 7,900,000. We will also assume a small initial number of infected individuals, let's say 10. The duration for this strain of flu was about three days giving us a recovery rate of 0.33 per day. We will assume an infection rate of 2.5×10^{-7} per day. The time interval will be one day.

We can now build the model in the spreadsheet and simulate it over time. We will need columns for each of the variables mentioned above plus cells to hold infection rate, recovery rate, and time interval. The initial spreadsheet setup should look like Figure 3-20.

Next, we will compute the new values for each of the populations by including their formulas (equations 3.4–3.6) in the spreadsheet. The example for the susceptible population is given in Figure 3-21. Its equation is B2-($G$2*B2*C2)*$I$2.

Subsequent simulation will produce a spreadsheet similar to that of Figure 3-22. Figure 3-22 shows 14 days worth of disease spread simulation.

	A	B	C	D	E	F	G	H
		Susceptible	Infected	Recoverd		Infection	Recovery	Time
1	Time	Population	Population	Population		Rate	Rate	Interval
2	0	7,900,000	10	0		2.50E-07	0.33	1

K ◄ ► ► Sheet1 Sheet2 Sheet3

Figure 3-20 Disease model spreadsheet setup

	A	B	C	D	E	F	G	H
		Susceptible	Infected	Recoverd		Infection	Recovery	Time
1	Time	Population	Population	Population		Rate	Rate	Interval
2	0	7,900,000	10	0		2.50E-07	0.33	1
3	1	7,899,980	26	3				

K ◄ ► ► Sheet1 Sheet2 Sheet3

Figure 3-21 Disease model at time one day

	A	B	C	D	E	F	G	H
		Susceptible	Infected	Recoverd		Infection	Recovery	Time
1	Time	Population	Population	Population		Rate	Rate	Interval
2	0	7,900,000	10	0		2.50E-07	0.33	1
3	1	7,899,980	26	3				
						Total		
4	2	7,899,928	70	12		Population		
5	3	7,899,790	185	35		7,900,010		
6	4	7,899,424	489	96				
7	5	7,898,458	1,294	258				
8	6	7,895,902	3,423	685				
9	7	7,889,144	9,051	1,815				
10	8	7,871,292	23,916	4,802				
11	9	7,824,229	63,087	12,694				
12	10	7,700,828	165,669	33,512				
13	11	7,381,881	429,946	88,183				
14	12	6,588,428	1,081,516	230,065				
15	13	4,807,055	2,505,989	586,966				
16	14	1,795,449	4,690,619	1,413,942				

K ◄ ► ► Sheet1 Sheet2 Sheet3

Figure 3-22 Disease model after 14 days

Figure 3-23 shows that graph of this data for that fourteen day period of time.

If you continued this problem for several more days you would see the number of susceptible individuals decrease to nearly zero resulting in the eventual dissipation of the disease. This type of simulation may be used to explore methods for combating the spread of disease by affecting parameters like the infection rate. You can explore different interventions and simulate the effect of the interventions before actual implementation. In this manner multiple intervention methods could be explored to see which is the most effective. This is a significant advantage of using simulation.

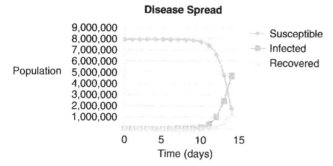

Figure 3-23   Disease spread over time

You will notice that this example is more complicated than the ball drop example. Here we have a system modeled by three dependent equations. Changes in one population affect the other populations. However, the spreadsheet simulation is capable of representing arbitrarily complex systems with these co-dependencies.

Example 4: Falling Ball Revisited

Here is one more example of using the spreadsheet method to construct a continuous simulation. With this example we will illustrate how inaccuracies may manifest themselves in this method. Recall the ball drop example. We left out one important aspect of the falling ball, the effect of air friction. Air friction tends to slow the acceleration of a falling object because of the *drag* created on the object. Drag is a function of the geometry of the object, its velocity, and a friction coefficient that accounts for how easily the air moves around the object. From a mathematical perspective the falling ball system can be represented by the following equation:

$$F = D - W \tag{3.11}$$

Here F represents the total force on the ball, W is the weight of the ball, and D is the drag produced by the air friction. We can rewrite equation 3.11 in terms of its primary components and obtain equation 3.12:

$$ma = kv^2 - mg \tag{3.12}$$

where m is the mass of the object, a is its acceleration, k is the drag coefficient, v is velocity of the object, and g is the gravitational constant. Thus we can calculate the acceleration of the falling body over time.

$$a = \frac{kv^2 - mg}{m} \tag{3.13}$$

You will note that the ball's acceleration is not constant like in the previous example, but changes as its velocity increases. We will set up the spreadsheet simulation taking into account this changing acceleration and look at the results of the simulation using two different time intervals: 0.1 seconds and 0.5 seconds. Figure 3-24 provides the simulation setup. The ball's mass will be 1 kilogram and the gravitational acceleration constant is as before.

The acceleration is decreasing due to the drag from air friction. The ball's velocity after 1 second is 8.60 meters per second compared to 9.8 meters per second in the previous example. From a model validation standpoint, accounting for drag provides for a more accurate representation of the real-world system.

Now let us see how our simulation is affected if we change the time interval to 0.5 seconds. Recall that this introduces more error in the system because of the approximation of the acceleration curve. Figure 3-25 shows the result of this simulation.

D2		f_x	=-($I$2*C2^2-$G$2*$H$2)/$G$2					
A	B	C	D	E	F	G	H	I
Time	Future Velocity	Current Velocity	Acceleration		Time Interval	Ball Mass	Gravity	Drag Coefficient
0	0.98	0	9.8		0.1	1	9.8	0.05
0.1	1.955198	0.98	9.75198					
0.2	2.916084	1.955198	9.608860039					
0.3	3.853566	2.916084	9.374822704					
0.4	4.759316	3.853566	9.057501348					
0.5	5.626061	4.759316	8.667445366					
0.6	6.447798	5.626061	8.217371912					
0.7	7.219928	6.447798	7.721294959					
0.8	7.939291	7.219928	7.193632249					
0.9	8.604129	7.939291	6.648383034					
1	9.213974	8.604129	6.098448069					

Figure 3-24   Spreadsheet simulation with 0.1 second time interval

A	B	C	D	E	F	G	H	I
Time	Future Velocity	Current Velocity	Acceleration		Time Interval	Ball Mass	Gravity	Drag Coefficient
0	4.9	0	9.8		0.5	1	9.8	0.05
0.5	9.19975	4.9	8.5995					
1	11.98386	9.19975	5.568229997					
1.5	13.29354	11.98386	2.619348985					
2	13.77558	13.29354	0.96409039					
2.5	13.93142	13.77558	0.311663328					
3	13.97931	13.93142	0.095781924					
3.5	13.99378	13.97931	0.028948354					
4	13.99813	13.99378	0.008703982					
4.5	13.99944	13.99813	0.002612954					
5	13.99983	13.99944	0.000784045					

Figure 3-25   Spreadsheet simulation with 0.5 second time interval

You will note that the velocity at 1 second is 9.199 meters per second compared to 8.60 meters per second even though we have not changed any other parameters in the system. This difference is strictly due to the error introduced from the approximation of the acceleration curve by the larger time interval. This example serves to illustrate one possible type of error that may result in employing this method for continuous simulation. The mathematics of calculus and differential equations provide a means to more accurately perform this type of simulation. That level of mathematical proficiency is beyond the scope of this textbook. However, what is important here is that relatively accurate simulations of continuous systems can be obtained using spreadsheet simulations.

CONCLUSION

This chapter discussed two types of simulation: discrete event and continuous. Both simulations add an aspect of *time* to the model. Discrete event simulation focuses on the occurrence of specific events to advance the model from one state to another state over time. Continuous simulation models behavior over a continuum of time independent of events that are occurring in the system.

Discrete event simulation depends on the chronological sequence of events that change the state of the system. These events are controlled by simulation time, the monitor that triggers a change in the system. This type of simulation allows us to study the behavior of a system in a compressed amount of time. Discrete event simulation functions by using an event list that represents what will happen in the future as well as controlling how the simulation changes its state with the passage of time.

Continuous simulation is used to simulate physical systems that involve mechanical, electrical, thermal, or hydraulic component as well as any continuous time-varying system. Equations are used to represent the state of the variables in the system. Difference equations are used to facilitate the use of algebra and spreadsheets to construct a continuous simulation.

KEY TERMS

discrete event	state variables	event list
simulation	simulation time	event generator
events	queuing system	continuous simulation
state of a system	random event	difference equations

REFERENCE

1. McCullough BD, Wilson B. On the accuracy of statistical procedures in Microsoft Excel 2000 and Excel XP. *Computational Statistics and Data Analysis* 2002;40(4):713–721.

Chapter 4

Queue Modeling and Simulation

Paul A. Fishwick and Hyungwook Park

INTRODUCTION

Queues are commonly found in most human-engineered systems where there exist one or more shared resources. Any system where the customer requests a service for a finite-capacity resource may be considered to be a **queuing system** [1]. Grocery stores, theme parks, and fast-food restaurants are well-known examples of queuing systems. Even a door or a toilet can be an example of a self-service queuing system. For example, McNickle used a queuing model to estimate the required number of toilets in New Zealand buildings based on the estimated number of people entering the buildings [2].

A queuing system can also be referred to as a *system of flow*. A new customer enters the queuing system and joins the queue, i.e., line, of customers, unless there is no queue and another customer who completes his service may exit the system at the same time. During the execution, a waiting line is formed in a system because the arrival time of each customer is not predictable, and the service time often exceeds customer inter-arrival times. A significant number of arrivals makes each customer wait in line longer than usual. **Queuing models** are constructed by a scientist or engineer to analyze the performance of a dynamic system where waiting can occur. In general, the goals of a queuing model are to minimize the average number of waiting customers in a queue and to predict the estimated number of facilities in a queuing system. The performance results of queuing model simulation are produced at the end of a simulation in the form of aggregate statistics.

Queuing theory was developed by Erlang in 1909 [3, 4].[1] Erlang worked for the Copenhagen Telephone Exchange as an engineer, and developed the tools to

---

[1] Erlang's other references are available at http://oldwww.com.dtu.dk/teletraffic/Erlang.html.

Principles of Modeling and Simulation: A Multidisciplinary Approach, Edited by John A. Sokolowski and Catherine M. Banks.
Copyright © 2009 John Wiley & Sons, Inc.

analyze and design a telecommunication system based on probability theory. In the 1950s and 1960s, queuing theory was associated with the fields of operation research and the performance analysis of time-sharing computer systems. In 1953, Kendall introduced the A/B/C type notation of a queuing system [5]. This notation is widely used for describing and classifying queuing systems. The Jackson network is regarded as the first significant achievement of the theory of queuing network in that each queue in a queuing network can be analyzed independently [6]. Little's Law defines the average number of customers to be equal to the arrival rate multiplied by the average time the customer spends in the system [7].

There are two approaches to estimating the performance and analysis of queuing systems: **analytical modeling** and **simulation** [8–10]. An analytical model is the abstraction of a system based on probability theory. The analytical model represents the description of a formal system consisting of equations used to estimate the performance of the system. Analytical modeling may be a better choice for a simple queuing system if we can easily derive the equations from the theory with a restricted set of assumptions. It is difficult to represent all situations in the real world because an analytical model requires assumptions such as infinite number of queue capacity and no bounds on the inter-arrival and service time. But, the queue size is limited and the bounds exist in the real systems. In some cases, a theory for the system equations is unknown or the algorithm for the equations is too complicated to be solved in closed-form. These restricted assumptions can result in inaccurate outputs.

A simulation is more flexible than an analytical model in that a simulation is not confined to working with formulas that have closed-form solutions. The complex arrival patterns and queue disciplines, which are often used as the assumptions, can be represented using a simulation model. Thus, simulation is often used to analyze the complex queuing systems in which analytical methods become intractable.

For the remainder of this chapter, we discuss queuing system model design and simulation by way of an example. The attributes, notation, and key statistics of a single-server queuing model are presented. We describe the sequential simulation of this queuing model followed by the graphical implementation of a single-server queuing model written in SimPack and Processing, which are two Java-based computer languages [11, 12].

ANALYTICAL SOLUTION

We show the analytical solution of a single-server queuing model assuming that the queue discipline is first-in first-out (FIFO), the customer who comes in first is served first. A queuing system can be modeled based on a Markov chain, which consists of discrete state spaces with the property that the next state depends only on the current state, and is independent of the previous state [1, 8]. Figure 4-1 shows the first five states of a Markov chain representing the states of a single-server queue.

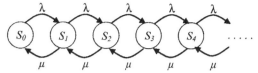

Figure 4-1   Markov chain

Each state is defined by the number of customers in the system. The number of customers in the system increases based on the arrival rate (λ) and decreases based on the service rate (μ). S_i denotes the ith state of a Markov chain, and the number i is the number of customers in the system. Traffic intensity or utilization (ρ) is defined by

$$\rho = \frac{\lambda}{\mu}, \text{ where } \lambda \text{ and } \mu \text{ refer to the arrival and service rate, respectively.} \quad (4.1)$$

Let p_i denote the probability that i customers exist in the system. The p_0, the probability for S_0 can be defined by

$$p_0 = 1 - \rho \quad (4.2)$$

In steady state, the probability for state k is represented by $\rho^k p_0$. The average number of customers in the system, $E(N)$ can be calculated by the sum of the number of customers multiplied by the probability in each state as follows.

$$E(N) = \sum_{i=0}^{\infty} \rho^i (1 - \rho) = \frac{\rho}{1 - \rho} \quad (4.3)$$

We can get the similar results with equation (4.3) from simulation; however the result takes longer to compute since we have to converge on the average by sampling over time.

Let us consider a second example involving two heterogeneous servers connected in parallel assuming that the service rates of two servers are different and the queue discipline is not FIFO but service in random order (SIRO). The number of states in a Markov chain exponentially increases as the number of servers increase. It is more difficult to derive the equations for probability than for the case of a single-server queuing model. Moreover, the different queue discipline requires the additional consideration when building an analytical model. If we assume the queue discipline is FIFO by reason that SIRO is too difficult to be modeled, the analytical model does not represent the real world exactly due to this assumption.

These restrictions do not present significant problems for discrete event simulation. The algorithm for the queue discipline and different service rate can be easily developed in the simulation model. We do not need to consider a lot of states for each queue, and the complicated computation is not a problem in discrete event simulation. In a discrete event simulation, the results are produced by running a simulation not by solving the equations [9].

QUEUING MODELS

Several checkout counters at the grocery store are parallel connections of single queues. If only one waiting line is formed for several checkout counters, each counter is regarded as one of servers for a single queue. A single-server queue, shown in Figure 4-2, is the simplest type of queuing model. However, if each checkout counter has its own waiting line, the queuing model is composed of several queues. Figure 4-3 shows the difference between parallel and single queues.

Attributes

A queuing model is described by its attributes: customer population, arrival and service pattern, queue discipline, queue capacity, and the number of servers [9]. A new customer from the calling population enters into the queuing model and waits

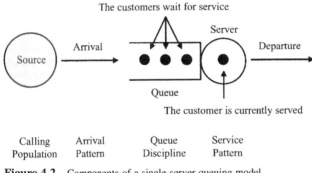

Figure 4-2 Components of a single-server queuing model

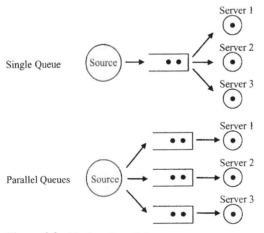

Figure 4-3 Single and parallel queues

for service in the queue. If the queue is empty and the server is idle, a new customer is immediately sent to the server for service, otherwise the customer remains in the queue joining the waiting line until the queue is empty and the server becomes idle. When a customer enters into the server, the status of the server becomes busy, not allowing any more arrivals to gain access to the server. After being served, a customer exits the system.

The *calling population*, which can be either finite or infinite, is defined as the pool of customers who possibly can request the service in the near future. If the size of the calling population is infinite, the arrival rate is not affected by others. But the arrival rate varies according to the number of customers who have arrived if the size of the calling population is finite and small. *Arrival* and *service patterns* are the two most important factors determining behaviors of queuing models. A queuing model may be **deterministic** or **stochastic**. For the stochastic case, new arrivals occur in a random pattern and their service time is obtained by probability distribution. The arrival and service rates, based on observation, are provided as the values of parameters for stochastic queuing models. The *arrival rate* is defined as the mean number of customers per unit time, and the *service rate* is defined by the capacity of the server in the queuing model. If the service rate is less than the arrival rate, the size of the queue will grow infinitely. The arrival rate must be less than the service rate in order to maintain a stable queuing system [1, 9]. The randomness of arrival and service patterns cause the length of waiting lines in the queue to vary.

Some customers in the real world may not stay in the queue upon arrival and leave the system. If the length of waiting line is too long or they find a shorter line, they may leave the queue. Balking, reneging, and jockeying are such examples of impatient customers [13]. Balking occurs when an arriving customer does not enter the queue due to the limited queue capacity. Reneging occurs when a customer leaves the queue after waiting in a queue upon arrival. Jockeying occurs when a customer decides to switch the queue for earlier service. The decision of balking is deterministic, whereas those of reneging and jockeying are considered as probabilistic.

When a server becomes idle, the next customer is selected among candidates from the queue. The selection of strategy from the queue is called *queue discipline*. Queue discipline is a scheduling algorithm to select the next customer from the queue. The common algorithms of queue discipline are first-in first-out (FIFO), last-in first-out (LIFO), service in random order (SIRO), and priority queue [9, 14]. The earlier arrived customer is usually selected from a queue in the real world, thus the most common queue discipline is FIFO. In a priority queue discipline, each arrival has its priority. The arrival that has the highest priority is chosen from queue among waiting customers. FASTPASS at Walt Disney World is a good example of a priority queue. The customer who has the FASTPASS is admitted earlier at the attraction than customers who do not have the FASTPASS.

The priority scheme may be either *preemptive* or *nonpreemptive*. In a preemptive scheme, the customer currently being served is placed back at the front of the queue if the incoming customer has the higher priority than the customer has who is currently being served. The displaced customer's service may be either restarted or resumed. In a nonpreemptive scheme, the continuous service is provided for the customer currently being served until it ends.

Notation

Kendall's notation, A/B/c/N/K, is used to concisely define a queue and its parameters. "A" and "B" represent the inter-arrival and service distribution, respectively; "c" represents the number of servers. "N" represents the queue capacity; "K" represents the size of the calling population. "D" (deterministic), "M" (Poisson), "G" (general), and "E_k" (Erlang) are used to represent "A" and "B". Usually the A/B/c notation is used when "N" and "K" are infinite. For example, M/M/1 represents a single server queuing model, and the inter-arrival and service time are exponentially distributed. The *queue discipline* is often added to describe the system. In this chapter, we will address only the M/M/1 type of queue.

Output

The purpose of building a queuing model and running a simulation is to obtain meaningful statistics such as the server performance. The notations used for statistics in this chapter are listed in Table 4-1, and the equations for key statistics are summarized in Table 4-2.

Table 4-1   Notations for Queuing Model Statistics

Notation	Description
ar_i	Arrival time for customer i
a_i	Inter-arrival time for customer i
$\bar{a}$	Average inter-arrival time
Λ	Arrival rate
T	Total simulation time
n	Number of arrived customers
s_i	Service time of ith customer
μ	Service rate
ss_i	Service start time of ith customer
d_i	Departure time of ith customer
$\overline{de}$	Mean delay time
$\bar{w}$	Mean residence time
ρ	Utilization
L	Number of customers in the system
B	System busy time
I	System idle time

Table 4-2 Equations for Key Queuing Model Statistics

Name	Equation	Description
Inter-arrival time	$a_i = ar_i - ar_{i-1}$	Interval between two consecutive arrivals
Mean inter-arrival time	$\bar{a} = \dfrac{\sum a_i}{n}$	Average inter-arrival time
Arrival rate	$\lambda = \dfrac{n}{T}$ $\lambda = \dfrac{1}{\bar{a}}$ (in the long run)	The number of arrivals at unit time
Mean service time	$\bar{s} = \dfrac{\sum s_i}{n}$	Average time for each customer to be served
Service rate	$\mu = \dfrac{1}{\bar{s}}$	Capability of server at unit time
Mean delay time	$\overline{de} = \dfrac{\sum (ss_i - ar_i)}{n}$	Average time for each customer to spend in a queue
Mean residence time	$\bar{w} = \dfrac{\sum (d_i - ar_i)}{n}$	Average time each customer stays in the system
System busy time	$B = \sum s_i$	Total service time of server
System idle time	$I = T - B$	Total idle time of server
System utilization	$\rho = \dfrac{B}{T}$	The proportion of the time in which the server is busy

Little's Law

The average number of customers (L) is equal to the arrival rate (λ) multiplied by the average time (w) the customer spends in the system: $L = \lambda w$. Little's law is meaningful in that the law holds regardless of any kind of the arrival and service distribution. Thus, Little's law does not require restricted assumptions for the types of arrival and service patterns.

Example: Grocery Store

Consider the following example of a grocery store with only one checkout counter. The customer waits for the cashier in a line. The arrival and service time for each customer are given in Table 4-3. Inter-arrival time and departure times can be

Table 4-3 First Eight Customers for a Grocery Store Server

Customer	Inter-arrival time	Arrival time	Service time	Departure time
1	---	2	4	6
2	1	3	2	8
3	2	5	3	11
4	3	8	1	12
5	5	13	2	15
6	1	14	2	17
7	1	15	1	18
8	2	17	3	21

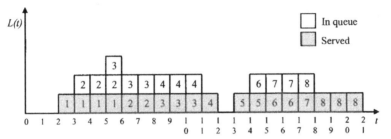

Figure 4-4 Number of customers over time

calculated from the arrival and service time. In a single-server queuing model, the customer waiting in the queue is not scheduled for service if another customer occupies the server. The start of service time is affected by the departure of the previous customer. Thus, the departure time (d_i) is defined as $d_i = \max (d_{i-1}, ar_i) + s_i$.

Figure 4-4 graphically illustrates the simulation results over time, showing the number of customers in the grocery store as time progresses. $L(t)$ denotes the number of customers in the system at time t.

The average customer $(\bar{L})$ at time t can be calculated from Figure 4-4 as follows. $\bar{L} = (1 + 2 + 2 + 3 + 2 + 2 + 2 + 2 + 2 + 1 + 1 + 2 + 2 + 2 + 2 + 1 + 1 + 1)/21 = 1.48$. We will now produce key summary statistics from the equations. The arrival rate (λ) is $8/21 = 0.38$. The mean residence time (w) is $(4 + 5 + 6 + 4 + 2 + 3 + 3 + 4)/8 = 3.88$. The average number of customers in the system can be calculated from Little's law: $L = \lambda w = 0.38 \times 3.88 = 1.48$. The result of Little's law is the same as that of simulation as shown in Figure 4-4.

Queuing Network

Queuing networks are classified into types: *open* and *closed* [8]. A *token* denotes any type of customer that requests service at the service facility. In an open queuing network, each token arrives at the system, based on the arrival rate, and leaves the system after being served. In the closed queuing network, the finite number of tokens is assigned, and each token moves between queues without leaving the system. The main difference between these two types of networks is that the open queuing network has new arrivals during simulation, whereas the closed queuing network does not have new arrivals. The number of tokens in the open queuing network at an instant of time is always different due to the arrival and departure rates, but the number of tokens in the system is always constant during the simulation of a closed queuing network.

SEQUENTIAL SIMULATION

In this section, we present the concept of discrete event simulation, and how a single-server queuing model is simulated sequentially by the concept of event scheduling. Discrete event simulation changes the state variables at a discrete time when the event occurs [15]. Thus, discrete event simulation has an advantage when the events occur irregularly compared to time-stepped simulation, state variables of which are computed at the end of the time step that is equally subdivided over the entire simulation time interval.

State variables have to be defined first when we build a model of discrete event simulation [14]. A state describes the quantifiable system attributes at any given point in time. For example, the number of customers in the system at time t is a state variable for a single-server queuing model. The state of the system is changed by the occurrence of the event, each of which is scheduled for execution and executed at a specific time. The arrival of the customer affects the number of customers in the system, suggesting that an arrival is a type of event. Arrival, service, and departure are all events for a single-server queuing model. A departure event is another arrival event for the next queue in the queuing network connected in tandem.

An *event scheduling* method is used along with a time-advanced algorithm. The simulation clock, the current simulated time, is set to 0 at the start of simulation and is updated when each event occurs [16]. Events occur randomly during the course of the simulation. A data structure, such as a linked list, heap, or tree, used for storing these events is called future event list (FEL). All events in the FEL are sorted in nondecreasing timestamp order when a new event is inserted. When the simulation starts, a head of the FEL, the event with the lowest timestamp, is extracted from the FEL for execution. The clock advances to the timestamp of the current event before the current event is executed.

Figure 4-5 shows the basic cycle for event scheduling. Three future events are stored into the FEL. When NEXT_EVENT is called, token #3 with timestamp 10 is extracted from the head of the FEL. The clock then advances into time 10. The event is executed at event routine 1; it creates a new event, event #2. Token #3 with event #2 is scheduled and inserted into the FEL. Token #3 is placed between token #2 and token #1 after comparing their timestamps. The event loop iterates to call NEXT_EVENT until the simulation ends.

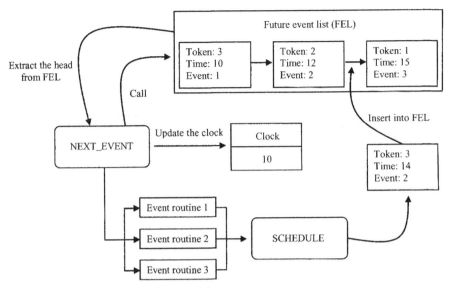

Figure 4-5   Cycle used for event scheduling[2]

```
public static void main()
     while (current_time <= simulation_time)
         Current_event = NEXT_EVENT ();
         Switch (event)
             Case 1: arrival
                 statements
             Case 2: service
                 statements
             Case 3: departure
                 statements
         end switch
     end while
end main

public event NEXT_EVENT ()
     nextEvent = the head of FEL;
     clock = the timestamp of nextEvent;
     return nextEvent;
end NextEvent
```

Figure 4-6   Algorithm of discrete event simulation for the single-server queuing model

Figure 4-6 shows the algorithm for the single-server queuing model, written in Java pseudo-code. For the stochastic queuing model, a new arrival event is created based on the last arrival time according to the probability distribution. The statements for each event can include interesting intermediate results of a queuing model. The current state of a queue and server can be drawn as a graphical representation, which will be illustrated in the following section.

---

[2] Fishwick, Paul A. (1995). *Simulation Model Design and Execution* [16], p. 57.

SIMPACK QUEUING IMPLEMENTATION

In the preceding section, we described the attributes, input, and output of a queuing model, and we specified how to build and run a single-server queuing model. We now present the programming source code of a single-server queue using SimPack and Processing. **SimPack** is a simulation toolkit which supports the construction of various types of models and executing the simulation.[3] SimPack is an application programming interface (API) based on an extension of the general-purpose programming language. C, C++, and Java versions of SimPack have been developed so far. SimPackJ, the Java version of SimPack, is used for scheduling and executing events whereas processing is used for representing the graphical results of the queuing model [17]. **Processing** is an open source programming language and environment that is used for representing images, animation, and interactions. Processing allows the Java program to be included inside of Processing. The result of SimPack is passed to Processing for graphical representation as the simulation progresses.

Figure 4-7 shows the definition of variables for the single-server queuing model. Token and event are declared. Class TOKEN represents a customer and has several attributes for each customer. Class SimEvent represents an event. Each event has four attributes: event time, event id, token, and its priority. Three events, ARRIVAL, SERVICE, and DEPARTURE, are defined along with the mean service time and inter-arrival rate.

Figure 4-8 shows the initialization for single-server queuing model in the *setup()* function. The setup() is used to define the initial environment and called when the program is started. The initialization of SimPack is included in setup(). The *Sim. init(minute()*60+second(), Const.LINKED)* is called to set up the initialization of discrete event simulation. When SimPack is used along with Processing, the current

```
import simpack.*;

public TOKEN token = new TOKEN();
public static SimEvent event = new SimEvent();

public static int ARRIVAL = 0;
public static int REQUEST_SERVICE = 1;
public static int DEPARTURE = 2;

public static int token_num=1;
public static double interArrival = 5.0;
public static double serviceTime = 5.0;
public static int next_time, server;
public float StartTime = minute()*60 + second();
public Customer [] Customers;
```

Figure 4-7 System variables for a single-server queue

[3] Documentation for SimPack is available at http://www.cise.ufl.edu/~fishwick/simpackj/doc/index.html.

```
// set up the graphical representation and simulation environment
void setup() {
  // initialize the discrete event simulation
  Sim.init(minute()*60+second(), Const.LINKED);
  server = Sim.create_facility("FACILITY", 1);
  Event_Arrivals(interArrival);

  // set up graphical representation
  size(600, 400);
  smooth();
  ellipseMode(CENTER);
  frameRate(30);
  PFont font;
  font = loadFont("Meta-Bold.vlw");
  textFont(font, 10);
}
```

Figure 4-8   setup() function

```
// Create and schedule new arrival event
void Event_Arrivals(double arrival_rate)
{
    int next_arrival = 0;
    SimEvent e1 = new SimEvent();
    e1.id = ARRIVAL;

    // generate interval based on mean inter-arrival time
    next_arrival = (int)Sim.expntl(arrival_rate) + 1;
    e1.token.attr[0] = (double) token_num;
    next_time += next_arrival;
    e1.token.attr[1] = (double) next_time;

    // schedule new arrival event
    Sim.schedule(e1, next_arrival);
    token_num++;
}
```

Figure 4-9   Generation method for new arrival

wall clock is usually set instead of the virtual clock for synchronizing with the physical clock for the graphical representation. The empty linked list is then created as the FEL. The facility with only a single server is created when *Sim.create_ facility("FACILITY",1)* is called.

As soon as the initialization is done for discrete event simulation, the first event is created and inserted into the FEL by calling the method *Event_Arrrival()*, shown in Figure 4-9. The next arrival event is created when the ARRIVAL event of the previous customer is executed. Inter-arrival time is determined by an exponential distribution based on the arrival rate. The simulation time of the newly created arrival event is the current clock value plus inter-arrival time.

The arrival event is scheduled and inserted into the FEL. The *Sim. schedule(SimEvent event, double time)* is used to schedule the event. When this method is called, the FEL is re-sorted in nondecreasing order according to its event time. The remainder of code in Figure 4-8 is needed to set up the graphical representation in Processing, such as the size of window and animation rate.

The *draw()* function, as shown in Figure 4-10, is called after the execution of the setup() function, and it iterates executing the lines of code inside the function

```
// iterate execution
void draw() {
  display();

  if (nts >0) {
    // extract the head from the FEL
    event = Sim.next_event(minute()*60+second(), Const.SYNC);
    int t_num = (int) event.token.attr[0];

    if(event.id == ARRIVAL) {
      event.id = SERVICE;
      Sim.schedule(event, 0.0);
      Customers[t_num].setArrival(Sim.state().
facility[server].queue_length, event.token.attr[1], t_num);
      display();
      event_arrivals(interArrival);
      Sim.update_arrivals();
    }

    else if(event.id == SERVICE) {
      if(Sim.request(server, event.token, 0) == Const.FREE) {
        int rn = (int)Sim.expntl(serviceTime) + 1;
        event.id = DEPARTURE;
        Sim.schedule(event, rn);
        Customers[t_num].moveToServer(t_num,
minute()*60+second(), rn);
        update(0);
        display();
      }
    }

    else if(event.id == DEPARTURE) {
      Sim.release(server, event.token);
      Customers[t_num].setDeparture();
      display();
      Sim.update_completions();
      nts--;
    }
  }

  else
    stats(); // print the summary statistics
}
```

Figure 4-10   draw() function

until the program is stopped. The code in the draw() function is used to display the updated graphical representation as the values of parameters change. The simulation repeats until the predefined number of customers has been served. An event with the lowest timestamp is extracted from the FEL when *Sim.next_event(minute()*60 +second(), SYNC)* is called. This method removes the head from the FEL, the event with the lowest timestamp. The ASYNC mode is used as a way to run the simulation in virtual time without synchronizing with the real-time (i.e., physical) clock, whereas synchronous execution mode (i.e., SYNC) will synchronize with the clock.

When the event is extracted from the FEL, the id of event is checked and sent to one of three event routines: *Arrival, Service,* and *Departure.*

- *Arrival:* Newly arrived token is scheduled for service and creates the next arrival based on exponential distribution.

- *Service: Sim.request(int facility, Token token, int priority)* is called to check whether or not the server is busy. If the server is idle, the service time is determined by the exponential distribution based on mean service. A token is scheduled for departure. If the server is busy, the event is stored into the queue, and requests the service again as soon as the previous token leaves the server. If *Sim.preempt(int facility, Token token, int priority)* is used instead of *Sim.request,* the priorities of requested and serviced tokens are compared for preemption.

- *Departure: Sim.release(int facility, Token token)* is called to release the token from the server. The token exits the system and the status of server is turned into the idle state. If there is a token which already requested the service but is blocked, the token is scheduled for service immediately. By calling *Sim.update_completions(),* the number of completion tokens is updated.

The methods, *Customers[].setArrival, moveToServer,* and *setDeparture* called during the event execution, are used to display the current location of tokens. The class Customer shown in Figure 4-11 defines those methods, and the display method is shown in Figure 4-12.

The *stats()* shown in Figure 4-13 is called to display the summary statistics when the simulation runs are complete.

Figures 4-14(a) and (b) show the intermediate results of the simulation. At time 7, customer #1 arrives at the system. Customer #1 enters the server immediately without waiting in the queue because the server is idle, as shown in Figure 4-14(a). Figure 4-14(b) shows the status that the bulk of customers arrives at the system. At time 32, customer #6 arrives at the system, but has to wait in the queue because other customers are in the system. Customer #6 starts its service at time 54 and will leave the system at time 58. Five customers are in the queue. At time 58, customer #7 will start its service.

Figure 4-15 shows the summary statistics for the single-server queuing model. Two hundred customers have been served during simulation. The system

```
// determine the location of each customer
class Customer {
  boolean display;
  int num, place, facility, arrivetime, servicetime;
  int servicestarted, servicestart, q_id, s_id;
  int QUEUE = 0, SERVER = 1, DONE = 2;

  Customer() {
    display = false;
    facility = QUEUE;
  }

  void setArrival(int qlength, int time, int id_) {
    num = qlength;
    arrivetime = time;
    place = QUEUE;
    display = true;
    q_id = id_;
  }

  void moveToServer(int id_, int t1, int t2) {
    place = SERVER;
    servicestarted = t1;
    servicetime = t2;
    servicestart = (minute()*60+second()-(int)StartTime);
    s_id = id_;
  }

  void setDeparture() {
    place = DONE;
    display = false;
  }

  void update(int f) {
    if(place == QUEUE && num > 0) {
      if(facility == f)
        num--;
    }
  }
}
```

Figure 4-11   Class customer

utilization is approximately 60%, and each customer waits in the queue for 3.9 on the average when we set the mean inter-arrival time to 5 and mean service time to 3.

This program can be extended with queues in tandem. Most fast-food restaurants have two facilities: one for ordering and one for pickup. The DEPARTURE event from the first server causes an ARRIVAL event for the second server instead of leaving the system. The request and release of the second server perform in the same manner that those of the first server do.

```
void display() {
    if(display) {
      fill(0, 255, 0);
      if(place == QUEUE) {
        ellipse(220 - 30 * (num%5), 200 - 90 * (num/5), 20, 20);
        text("#" + q_id, 215 - 30 * (num%5), 230 - 90 * (num/5));
        text((int) arrivetime, 215 - 30*(num%5), 245 - 90*(num/5));
      }

      else if(place == SERVER) {
        ellipse(300, 200, 20, 20);
        text("Customer : #" + s_id, 380, 120);
        text("Arrival Time : " + (float) arrivetime, 380, 140);
        text("Service Start: " + (float) servicestart, 380, 160);
        text("Service Time : " + (servicetime - (float)
(minute()*60+second()-servicestarted)), 380, 180);
      }
    }
  }
```

Figure 4-12 Method display() in class customer

```
void stats() {
    background(255);
    textSize(16);
    noFill();
    stroke(0, 0, 255);
    rect(25, 20, 250, 30);
    fill(0, 0, 255);
    text("SimPackJS SIMULATION REPORT",35, 40);

    // call the statistics method in SimPack
    Sim.report_stats(StartTime);
    text("Total Simulation Time: "+ Sim.total_sim_time,30, 80);
    text("Total System Arrivals: "+ Sim.arrivals,30, 100);
    text("Total System Completions: "+ Sim.completions,30, 120);
    text("System Wide Statistics",30, 200);
    line(30,210, 192, 210);
    text("System Utilization: "+100.0*Sim.total_utilization,30,
240);
    text("Arrival Rate: "+ Sim.arrival_rate ,30, 260);
    text("Throughput: "+ Sim.throughput,30, 280);
    text("Mean Service Time per Token: "+
Sim.mean_service_time,30,300);
    text("Mean # of Tokens in System: "+
Sim.mean_num_tokens,30,320);
    text("Mean Residence Time for each Token: "+
Sim.mean_residence_time,30, 340);
  }
```

Figure 4-13 Method stats()

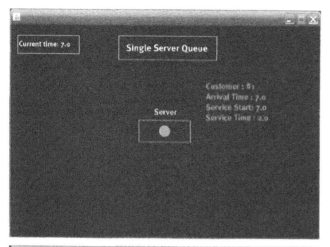

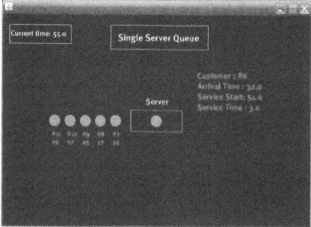

Figure 4-14 (a) One customer is being served (b) A queue forms behind the customer being served

PARALLEL SIMULATION

We presented the sequential simulation method of a queuing model. The event size of the queuing network is generally fine-grain. For a small queuing network, a sequential simulation does not take a long time to simulate. For large-scale queuing network models however, the number of events proportionally increases as the number of nodes increases. The events are stored and processed by a single event list due to memory capacity and processor capability. Moreover, a single processor is responsible for executing the event as well. **Parallel simulation** is a solution for large-scale queuing network models. Synchronization is required to run a parallel simulation because the produced results are expected to be exactly the same as those produced by sequential simulation. More information on parallel simulation is provided in [18].

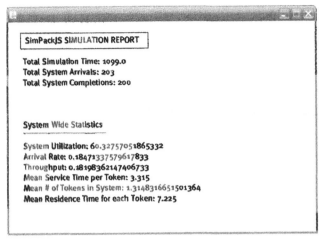

Figure 4-15   Summary statistics produced by SimPack

CONCLUSION

We introduced the queuing model and its simulation. Queuing systems are commonly found where the resources are shared. The analytical modeling and simulation are two approaches used for analyzing queuing systems. An analytic model describes the formal system defined by a set of equations, and is used when the model is simple enough and the closed-form solutions are available with a restricted set of assumptions. A simulation is used to analyze the complex queuing systems where an analytic modeling approach is not feasible. The results of the simulation are produced by executing a queuing model through numerical means, as opposed to performing symbolic computation to solve for closed-form solutions. We presented the example of an analytical solution and its restrictions compared to simulation.

Queuing models are described by several attributes. The calling population and queue discipline have an effect on the arrival and service rates. The arrival and service patterns are the most important attributes on the behaviors of queuing models, and the arrival rate must be less than the service rate to keep a queuing system stable. The key statistics are produced at the end of a simulation based on the arrival and service rate.

Queuing models are simulated using discrete event simulation. State variables and events are defined to describe the system and to change the state of the system over time. The events are stored in the FEL and executed in their timestamp order. The simulation clock is updated when each event occurs. A newly created event during the simulation is scheduled and inserted into the FEL according to its timestamp.

We presented the programming source code of a single-server queue developed in SimPack and Processing. SimPack along with Processing shows us the progresses of the simulation as a graphical representation over time. The concept of parallel simulation is very briefly summarized for those who need fast execution when simulating large-scale queuing network models.

Queuing models will continue to play a major role for most engineering disciplines wherever there exists humans, jobs, or tasks that must wait for a shared resource. These types of models tend not to occur in natural systems primarily because the nature of queuing is handled through other means involving transport equations. As queuing models grow in size and complexity, it is possible to treat them in a "continuous flow" manner using partial differential equations; however, for the majority of systems, the methods discussed in this chapter should be sufficient to initiate basic computational experiments with queues.

KEY TERMS

queues	simulation	SimPack
queuing system	deterministic	Processing
queuing models	stochastic	parallel simulation
analytical modeling	Kendall's notation	

FURTHER READING

BANKS J, CARSON JS, NELSON BL, NICOL DM. *Discrete-Event System Simulation*. Upper Saddle River, NJ: Prentice Hall; 2005.
LAW AM, KELTON WD. *Simulation, Modeling, and Analysis*. 4th Ed. New York: McGraw-Hill; 2006.
LEEMIS LM, PARK SK. *Discrete-Event Simulation: A First Course*. Upper Saddle River, NJ: Prentice Hall; 2006.
ROSS SM. *Introduction to Probability Models*. New York: Academic Press; 2006.

REFERENCES

1. KLEINROCK L. *Queueing Systems Volume 1: Theory*. New York: Wiley-Interscience; 1975.
2. MCNICKLE D. Queuing for toilets. *OR Insight* 1998; April–June.
3. COOPER RB. Queuing theory. In: Levy B, ed. Proceedings of the ACM '81 Conference; 1981 New York; pp. 119–122.
4. BROCKMEYER E, HALSTROM HL, JENSEN A. The life and works of A. K. Erlang. *Transactions of the Danish Academy of Technical Sciences* 1948;2.
5. KENDALL DG. Stochastic processes occurring in the theory of queues and their analysis by the method of imbedded Markov chains. *Annals of Mathematical Statistics* 1953;24:338–354.
6. JACKSON JR. Networks of waiting lines. *Operation Research* 1957;5:518–521.
7. LITTLE JDC. A proof of the queuing formula $L = \lambda W$. *Operations Research* 1961;9:383–387.
8. BOLCH G, GREINER S, DE MEER H, TRIVEDI KS. *Queuing Networks and Markov Chains: Modeling and Performance Evaluation with Computer Science Applications*. New York: Wiley-Interscience; 2006.

9. BANKS J, CARSON JS, NELSON BL, NICOL DM. *Discrete-Event System Simulation*, 4th Ed. Upper Saddle River, NJ: Prentice Hall; 2005.

10. SAUER CH, MACNAIR EA. *Simulation of Computer Communication Systems*. Upper Saddle River, NJ: Prentice Hall; 1983.

11. FISHWICK PA. Simpack: Getting started with simulation programming in C and C++. In: Proceedings of the 24th conference on Winter simulation; 1992. New York: ACM Press; 1992. pp. 154–162.

12. Available at http://www.processing.org/. Accessed 2008 March 15.

13. CHUNG CA. *Simulation Modeling Handbook: A Practical Approach*. Boca Raton, FL: CRC Press; 2003.

14. LEEMIS LM, PARK SK. *Discrete-Event Simulation: A First Course*. Upper Saddle River, NJ: Prentice Hall; 2006.

15. LAW AM, KELTON WD. *Simulation, Modeling, and Analysis*, 4th Ed. New York: McGraw-Hill; 2006.

16. FISHWICK PA. *Simulation Model Design and Execution: Building Digital Worlds*. Upper Saddle River, NJ: Prentice Hall; 1995.

17. PARK M, FISHWICK PA. Simpackj/s: A web-oriented toolkit for discrete event simulation. In: Proceedings of Enabling Technology for Simulation Science, Part of SPIE Aerosense '02 Conference; 2002. pp. 348–358.

18. FUJIMOTO RM. *Parallel and Distribution Simulation Systems*. New York: Wiley-Interscience; 2000.

Chapter 5

Human Interaction with Simulations

Michael P. Bailey and Ahmed K. Noor

INTRODUCTION

Let's review what we have discussed thus far. In Chapter 1 we learned the basics: what is M&S, what are models, what are simulations. Chapter 2 emphasized the role of modeling and simulation problem solving for gaining insight into complex issues. Chapters 3 and 4 focused on simulation as models that vary over time with an emphasis on queue modeling and simulation. This chapter introduces human interactions with simulation. It is divided into two subsections that were written by individuals who are expert in this area of M&S research. First, we will look into *Simulation and Data Dependency*, by Michael P. Bailey, Ph.D., followed by a discussion on *Visual Representation*, written by Ahmed K. Noor, Ph.D.

SIMULATION AND DATA DEPENDENCY[1]

Michael P. Bailey[1]

This section is dedicated to exploring the uses of simulation in support of analysis. To frame the discussion and to identify the unique properties of the activity, the foundation of analysis needs to be discussed. A practical statement of the **scientific method** is …

> *The formulation and evaluation of a hypothesis using empirical evidence.*

---

[1] The author would like to acknowledge the assistance of Mr. Kevin Hankins in constructing some of the graphics and lending his sharp intellect to this effort.

Principles of Modeling and Simulation: A Multidisciplinary Approach, Edited by John A. Sokolowski and Catherine M. Banks.

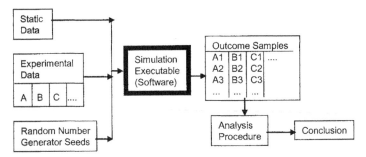

Figure 5-1   Basic simulation-supported experiment

The modern rendition of the scientific method in industrial, military, governmental, and social science pursuits is well-described in the seminal book, *Methods of Operations Research* by Morse and Kimball (1951), which is the foundation of modern operations research [1].

When we consider using simulation in pursuit of a scientific investigation, our approach is to design and execute a controlled experiment. The structure of this experiment is shown in Figure 5-1.

Here are some things to consider when using simulation in pursuit of scientific investigation:

- In pursuit of knowledge about the system, the analyst/modeler who does all of the work involved in constructing the simulation software will learn a great deal. The steps involved in this development are extremely rewarding and informative.

- The modeling and software development for an analysis-supporting simulation are affected by the targeted use of the simulation in the experiment, and detail is added where it helps differentiate between the alternatives.

- The accuracy of the static data is decisive in the success of the investigation, though the time and effort dedicated to data development is often much less than that dedicated to developing the software.

- The system under investigation is assumed to have important dynamics that are **stochastic** (governed by effects that are modeled probabilistically). Hence, our experiment will involve producing samples of the outcome through statistically independent replications.

- We have narrowed the scope of scientific investigation to the pursuit of a single metric, and our goal is to determine the effect of changing the experimental input data on this outcome metric.

For example, many colleges and universities are housed on military bases. Such is the case for the Naval Postgraduate School in Monterey, California. As a result, all students, professors, administrators, and other personnel are required to go through a security checkpoint to gain access to the school. This is necessary for security and safety, but can also be very inconvenient if the daily delay is too long.

Operations Research Department Chairman Pierre Lafayette often complains about his long wait at the gate. New Operations Research Assistant Professor Willie Tenure has decided to use his skills in analysis, simulation, and experimentation to help his chairman.

So, Willie spends some time studying the operations at the security gate. He's decided that the following modeling characteristics and assumptions describe the gate fairly well:

- There's no appreciable incoming traffic prior to 6:30 AM
- At 6:30 AM the traffic picks up to about one car every five minutes
- At 7:00 AM the traffic intensifies to about 1.2 cars per minute
- All of the cars patiently wait to get through the gate; no one leaves, jumps in line, or gets special treatment.
- The guard takes a minimum of 30 seconds to check identification, look through the car, and check its window sticker. He usually takes about 45 seconds to do this, and he never takes more than two minutes.
- Pierre appears like clockwork at 7:30 AM each day

Willie constructs a simple simulation model in the specialized computer program *Arena* [2]. This simulation can also be constructed using the spreadsheet methodology of Chapter 3.

The Arena flowchart, shown in Figure 5-2, is easy to interpret. The three creation nodes (three sideways pentagons) create cars that enter the gate queue. The top one generates the pre-7:00 cars at a rate of one per five minutes, the middle create node adds the post-7:00 cars at a rate of one car per minute. Thus, the combined traffic is the required 1.2 cars per minute. The third create node only generates one car, Pierre, at exactly 7:30.

All of the cars are blended into one input stream and they are each stamped with their arrival time and enqueued at the gate. They await their turn at the gate in the order of their arrival. The gate and guard, depicted with the rectangular node, process each car in turn. Upon completion, the simulation records how long the car waited, and disposes of the car.

Two aspects of the gate operations display **randomness**—the time between arriving cars (the inter-arrival time) and the time the guard takes with each car (the service time). Thus, for inter-arrival times it is common practice to model arrivals of this type (independently acting cars coming from a near-infinite source) as separated by a random time following the exponential distribution. For service times when minimum (30 seconds), maximum (2 minutes), and most likely (45 seconds) values are known, a simple and common choice uses the triangular distribution as the model.

These decisions on the choice of probability model for different system dynamics are based on specialized probability theory, some of which is pretty sophisticated.

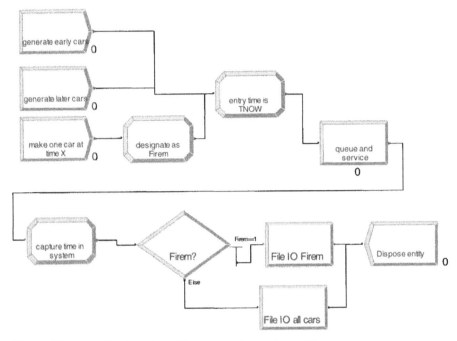

Figure 5-2 Arena flowchart view of the gate security queuing model

A good starting point for exploring input analysis can be found in a text published by Averill Law and David Kelton, *Simulation Modeling and Analysis* [3]. Arena provides easy-to-access support for generating a wide variety of stochastic waiting times and inter-arrival intervals.

Simulation Runs

Willie Tenure ran his simulation through to Pierre's arrival, wait, and completion of his security check. During this run

- six cars appeared between 6:30 AM and 7:00 AM
- 32 cars arrived between 7:00 AM and 7:30 AM
- when Pierre arrived, there were seven cars waiting ahead of him
- Pierre waited 10.05 minutes to get through, including spending 0.91 minutes having his own identification and sticker checked

When Willie Tenure makes a second run, the values 6, 32, 7, 10.05, and 0.91 change to 9, 26, 6, 14.49, and 1.21; 10.05 and 14.49 are two independent samples of Pierre's wait at the gate. Table 5-1 shows 30 such samples generated with Arena. We will use all of this data to form a statistical picture of the time Pierre has to wait.

Table 5-1   Waiting Times of Pierre Lafayette at the Naval
Postgraduate School

10.050491	14.913239	16.922335
14.493382	12.134188	2.806307
13.098571	10.678704	7.074584
17.03254	26.322664	10.232526
16.727695	17.054722	16.491132
8.333908	8.779273	14.311304
8.303289	9.116717	2.291182
0.9965	10.727221	10.905171
22.535214	7.365525	17.332115
8.834664	19.284139	19.267145

Simulations used for analysis come in a variety of forms, but there are two pre-dominant ones.

Networks of Queues

Entities are generated and routed to a processing station where they await one or more resources or servers. Upon completion, the entity may be routed to another station, and the server turns its attentions to the next entity in line. Telecommunications packets, hamburger buyers, and parts in a factory are all types of entities modeled this way. There exist some well-established simulation packages that enable easy model coding when this particular form is appropriate, including Arena, Extend, and COMNET.

Roaming Entities

A heterogeneous collection of entities move about a space, a network, or a plane according to goals the entities have. When they come in proximity of each other, they interact. This interaction could involve transfer of information, rendering of assistance, or combat. Many of the simulations in military operations research use this type of simulation, but there are also great examples in population move-ments, automobile traffic, and even telecommunications. In these latter examples, the space is structured as a network. Recent developments in agent-based simula-tion have spurred the development of packages like Mana and Pythagoras [4].

Statistical Treatment of Outcomes

All of the analysis procedures we will discuss here are based on the assumption that replications of our simulation have provided numerous, statistically independent, identically distributed data. In some important cases this assumption turns out to be false, and we'll address those by manipulating and excluding data until we've got data with these properties (at least approximately!).

We're going to use the **Lemma-Theorem-Corollary structure** common in higher-level mathematics texts for convenience. Don't be intimidated, the meaning of these terms is simple: lemmas are building blocks for a theorem; the theorem is the main important fact (usually very general) we're trying to establish; and corollaries are meaningful facts derived straight from the theorem. First, let's establish two useful facts for expected values and variances.

Lemma 1 Let w_1, w_2, ... , w_n be independent and identically distributed samples with expected value $E[w] = \mu$ and variance $VAR[w] = \sigma^2$, then

$$E[cw] = c\mu$$
$$VAR[cw] = c^2\sigma^2$$

(5.1)

Next, we need to establish some facts about sums of independent, identically distributed (iid) data.

Lemma 2

$$E\left[\sum_{i=1}^{n} w_i\right] = n\mu$$

$$VAR\left[\sum_{i=1}^{n} w_i\right] = n\sigma^2.$$

(5.2)

All our methods start with application of these two facts, and the **Central Limit Theorem** (CLT), which we state here:

Central Limit Theorem. Let w_1, w_2, ... , w_n be independent and identically distributed (iid) samples with expected value $E[w] = \mu$ and variance $VAR[w] = \sigma^2$. Then if n is large enough

$$\sum_{i=1}^{n} w_i \sim N(n\mu, n\sigma^2).$$

(5.3)

The notation $N(\mu, \sigma^2)$ is used to denote the normal distribution with mean μ and variance σ^2. A special case of this distribution is the standard normal, $N(0, 1)$, with mean 0 and variance (and standard deviation) 1. All of this allows us to construct the large-sample confidence interval for the mean of iid data.

Corollary 1

$$E\left[\sum_{i=1}^{n} \frac{w_i}{n} - \mu\right] = 0,$$

$$VAR\left[\frac{\sum_{i=1}^{n} \frac{w_i}{n} - \mu}{\frac{\sigma}{\sqrt{n}}}\right] = 1,$$

(5.4)

so, $\dfrac{\sum_{i=1}^{n} \frac{w_i}{n} - \mu}{\frac{\sigma}{\sqrt{n}}} \sim N(0, 1).$

We left the expression for the mean of the w_i's in its complex form, so the connection between the lemmas, the theorem, and the corollary is obvious. From here on, we'll use the notation $\bar{w}$ for the mean of the w_i's. Let Z be a $N(0, 1)$ random variable, then $z_{\alpha/2}$ be a standard normal p-tile for $\alpha/2$, defined as satisfying the equation

$$P[Z \le z_{\alpha/2}] = \alpha/2, \text{ or}$$
$$P[z_{\alpha/2} \le Z \le z_{1-\alpha/2}] = 1 - \alpha \tag{5.5}$$

Since the complicated final expression in the corollary is also a standard normal (the corollary says so) we can say …

$$P\left[z_{\alpha/2} \le \frac{\bar{w} - \mu}{\dfrac{\sigma}{\sqrt{n}}} \le z_{1-\alpha/2} \right] = 1 - \alpha, \text{ or}$$

$$P\left[\mu - z_{\alpha/2}\frac{\sigma}{\sqrt{n}} \le \bar{w} \le \mu - z_{1-\alpha/2}\frac{\sigma}{\sqrt{n}} \right] = 1 - \alpha, \tag{5.6}$$

Standard normal p-tile values are common to most statistical packages, modern spreadsheets with data analysis plug-ins, and all elementary statistics texts. Note that equation 5.6 is a probability statement on $\bar{w}$ not on μ—μ is an unknown constant not a random variable. The correct interpretation of the interval in equation 5.6 is that this interval *covers the unknown value of μ with probability 1 − α*.

Example We ran the simulation model for the security gate 30 separate times, and recorded Pierre's waiting time. These times are shown in Table 5-1 and depicted in Figure 5-3. The mean for this data is 12.485. Using the formula from equation 5.6, except substituting the estimate s = 5.858 for the unknown constant σ, we calculate the confidence interval for Pierre's waiting time as [10.384, 14.577]. We are 95% confident that this interval, calculated from observations, covers the true expected duration of Pierre's wait.

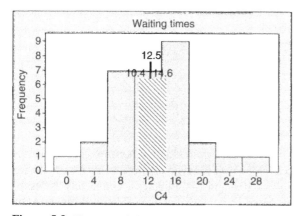

Figure 5-3 Histogram of simulated waiting times with the 95% confidence interval of the mean overlayed

How Many Replications?

Typically, when statistics are computed, there is some cost to collecting the data. Unlike data collected from surveys, measurememts, or field observations, simulation data is essentially free—you can have as many replications as you want for no additional cost. This luxury comes with an implied question: How many data are enough?

You have enough data when ... the *large enough* condition of the Central Limit Theorem is satisfied and the confidence interval constructed is sufficiently tight. Let n denote the number of samples w_i we will collect from the simulation. The Central Limit Theorem's sample size assumption is usually satisfied for $n > 30$ to 50. This guideline is good as long as the data are somewhat symmetric and unimodal (as opposed to multimodal—having several significant humps in the histogram). Plotting a histogram is a recommended check for this condition.

The tightness of the calculated confidence interval is driven by its radius or half-width r,

$$r = \frac{z_{\alpha/2}\sigma}{\sqrt{n}}. \tag{5.7}$$

Since $z_{\alpha/2}\sigma$ is a fixed quantity, r will shrink as n gets large. If we wanted to estimate the value of E[W] within a tolerance of $\pm\delta$ (δ can be stated in absolute terms, or as a percentage of E[W]), we have

$$\delta > \frac{z_{\alpha/2}\sigma}{\sqrt{n}}, \tag{5.8}$$

implying

$$n_{min} > \left(\frac{z_{\alpha/2}\sigma}{\delta}\right)^2. \tag{5.9}$$

So, as δ gets smaller, the necessary n increases and the resulting confidence interval shrinks as a result. To get a sufficiently tight confidence interval of radius δ ...

1. make an exploratory simulation run of ~50 samples;
2. estimate σ;
3. calculate n_{min};
4. make a production run with n_{min} replications;
5. compute the resulting confidence interval.

Example Willie Tenure is a typical assistant professor. His security gate simulation is elegant, his statistical treatment of the data is beyond reproach, and his conclusion that Pierre's waiting time is between 10.4 and 14.6 minutes is completely useless to Pierre. A more accurate estimate based on a tighter confidence interval supported by more simulation runs is needed. Pierre shouldn't be satisfied with an answer with more than 1.0 minutes of uncertainty. Using equation 5.9, with $\delta = 0.5$ minutes, we get

$$n_{\min} > \left(\frac{z_{\alpha/2}\sigma}{\delta}\right)^2 \approx \left(\frac{(1.96)(5.86)}{0.5}\right)^2 \approx 527.4 \tag{5.10}$$

Rerunning the simulation 600 replications, we observe that the mean did indeed shift slightly higher, 12.38 (you wouldn't expect it to remain the same) and our new confidence interval is [11.9, 12.8], less than one minute wide.

Factors, Levels, and Measures

Let's refocus our attention on the scientific method. In any analysis, our investigation is necessarily narrowed to a small number of important questions. Part of the cost and benefit of analytic rigor is this narrowing of scope. We need to determine which variables will be our experimental variables (the A, B, C from Figure 5-1), we need to isolate what is really important in terms of system performance, and we need to develop a formal hypothesis that states the relationship of the input variables to the outcome.

The outcome of the system is called its **measure of performance**. The properties of an appropriate measure are that it must be:

- quantifiable (real-values, integer, etc.)
- accessible and measurable (not a problem for a simulation-based experiment)
- aligned with the goals of the experiment
- intuitively interpretable and meaningful

On the input side, we must identify what important parameter(s) will be static, and which will be experimental. Often, this is an obvious choice, but some considerations merit reflection. **Static variables** are static; they do not change over the course of the experiment. **Experimental variables** (let's start with just one) are often called *factors* in the language of experimental designs. Choices for the values of a factor are called its *levels*. For the investigation to be of value, the experimental factors must be things that a decision maker can change. They must be actionable, and their levels should be realistic. If there is uncertainty about a parameter value that is not grounds for making it experimental. If there are some interesting alternatives for static variables, these should be treated as cases.

Example In our ongoing security gate question, the obvious measure of system performance is the time Pierre Lafayette must wait at the gate. The experimental factor that is under his control and that can be manipulated to reduce his wait is his chosen arrival time currently set at 7:30 AM. The traffic parameters for all of the other cars are static (at least for now!)

Example Mel Fencejumper decides to go to the playoff game involving his favorite NFL team, the Green Bay Packers. He doesn't yet know which of the NFC wildcard teams will be facing the Packers, but he does know that the opponent, either the Seattle Seahawks or the Washington Redskins, will have a dramatic effect on the price for aftermarket tickets (see Table 5-2).

Mel intends to approach the stadium on game day and buy a ticket from one of these sellers, but his wife has set a limit for him of X. We want to know what the effect of X is on Mel's experience.

Mel's measure of performance is clearly the price he will pay for his ticket. In this experiment, we'll ignore other factors like the location of his seat, the number of aftermarket sellers he has to approach before he can buy a ticket, or the time he arrives at the stadium. Also clear, the opposing team is not an experimental variable; Mel gets no say in who plays the Packers. The price limit, X, is the only factor in our experiment.

One alternative is to treat the options for opposition as two cases. Let's consider levels as shown in Table 5-3. For each case, we generate 15 random ticket prices according to the uniform distribution with the parameters from Table 5-2, and enforce Mrs. Fencejumper's limit at $200.

Table 5-2 Minimum and Maximum Ticket Prices for the Upcoming Packers Playoff Game

Team	Min	Max
Redskins	150	500
Seahawks	130	300

Table 5-3 Simulated Ticket Prices

	Redskins	Seahawks
1	444.03	208.13
2	165.05	215.07
3	395.25	104.41
4	202.62	294.04
5	278.39	287.67
6	157.22	112.95
7	377.78	237.6
8	493.41	105.07
9	444.25	149.81
10	384.7	259.9
11	350.93	139.15
12	264.89	145.38
13	233.85	277.07
14	423.16	283.87
15	305.48	198.83
AVG(200)	161.13	136.51

The AVG(200) formula calculates the average of all of the samples that are less than $200. Thus, if the Redskins are coming to town, the average ticket cost is going to be about $160, while if the Seahawks are the opposition, the average cost will be only $136.50.

While this example could have been worked analytically using conditional probability distributions, we chose to use simulation to make a point—*the results from a simulation are not exact, and the accuracy of our answers is dependent on the size of our sample.* The true average ticket prices, given the $200 limit, are easily computed as $175 for the Redskins, $150 for the Seahawks. If we did this experiment over and over, those are the averages we would eventually converge upon.

Supporting a Real Experiment—Analyzing Differences

Building a **confidence interval** for a Measure of Performance is a good exercise that allows an analyst to communicate the magnitude, as well as the uncertainty, of the measurement. However, this doesn't really meet the objectives of the experiment from Figure 5-1.

In Figure 5-4 we see the 30-sample confidence intervals for the expected waiting time corresponding to three alternative arrival times for Pierre Lafayette. You can clearly see they are different, that the longer Pierre sleeps, the longer he can expect to wait at the gate. But what can you say beyond that? Can you say that he'll save five minutes waiting by arriving at 7:15 AM as opposed to 7:30 AM? The center points of these confidence intervals are more than 5.0 minutes apart, but the top of the 7:15 AM interval is only a little over 2.0 minutes below the bottom of the 7:30 AM interval. How do we handle this comparison elegantly? The answer is to construct many

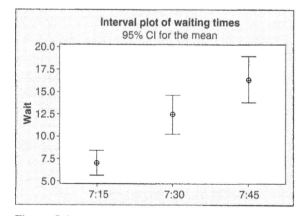

Figure 5-4   Confidence intervals for the expected waiting time for three different arrival time choices

samples of the difference between the two waiting times, and treat these samples statistically.

Example Well, Willie Tenure has again proven his profound ignorance of the real problem. Telling Pierre his current average wait, no matter how accurately, won't score points with Pierre. What Pierre needs are options—other arrival times when he can significantly reduce his wait without losing too much beauty sleep, or he needs to be told that there just isn't any time after 7:00 AM when his wait will decrease significantly.

Let w_1, w_2, ... , w_{30} denote the waiting times for a 7:30 AM arrival, and let y_1, y_2, ... , y_{30} denote the waiting times for a 7:15 AM arrival. Let's form the samples of the difference in the two waiting times ...

$$d_i = w_i - y_i, \quad i = 1, 2, \ldots, 30. \tag{5.11}$$

Thus, the d_i's are 30 iid samples of the difference between the two waiting times. We can do all of the things one does with iid data, including calculating a mean and confidence interval. The average waiting time savings if Pierre arrives at 7:15 AM is 5.03 minutes, and the 95% confidence interval if [3.38, 6.68]. If this 3-minute interval is too sloppy, we can use equation 5.9 to calculate the sample size required to satisfy any accuracy standard. By the way, the waiting time penalty for sleeping in and arriving at 7:45 AM has average 4.14 minutes with 95% confidence interval [1.66, 6.23]. Now Pierre knows the cost of his beauty sleep in terms of extra waiting at the security gate, and Willie's tenure case is sure to be successful.

Analyze Queues with Caution

In our ongoing example we started a replication of the simulation with an empty waiting line and light traffic intensity. After a delay the arrival rate increased, there was another delay, and then we introduced Pierre's car and captured his waiting time. Then we stopped the simulation. We captured only one datum per replication. If our goal was to capture the average car's waiting time, we might be tempted to just capture every car's waiting time as they exited the security area, and average those times. There are some problems with this:

- The early cars, due to the low arrival rate and the fact that it's likely that there are few cars in front of them, are likely to have low waiting times. Cars later in the day would expect to have larger waiting times. Probabilistically, the early cars and the late cars are *not identically distributed.*

- Suppose you are in car n, and car n-1 enjoys a shorter-than-average wait because the queue he arrived at was nearly empty. This tells you that your car will also, more than likely, enjoy a shorter-than-average wait. Probabilistically, cars n-1 and n have waiting times that are *not independent.*

Thus, the assumptions of the Central Limit Theorem are violated, and the methods we have explored in this chapter are not valid. What's worse, if you were

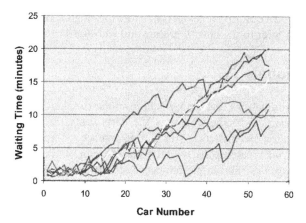

Figure 5-5 Sample paths of sequential waiting times for nine independent replications, demonstrating that these times are not iid

to naively employ these methods, you would likely produce a confidence interval that grossly overstates the accuracy of your estimate of the expected waiting time. Figure 5-5 shows nine replications of the security gate simulation for the first 55 cars. You can readily see that the early waiting times are smaller, and that these times are obviously not independent of one another.

Simulation is a very powerful tool for analysts, providing an inexhaustible source of data to support the Scientific Method. The researcher or professional in this area gets to work in probabilistic modeling, statistics, and computer science. The opportunity to expand your expertise in this eclectic set of techniques, methods, and theories is great. This makes simulation for analysis a challenging and rewarding pursuit, and makes serious professionals in this field sought after in industry, government, and academia.

VISUAL REPRESENTATION

Ahmed K. Noor

Visual representations are reshaping the exploration, analysis, and dissemination of simulation data, and promise to revolutionize knowledge discovery and innovation. In this subsection a brief discussion is presented of some selected aspects of visual representations. These include: the evolution of and rationale for visual representations; four categories of visual representations; virtual reality and virtual worlds; current focus and challenges of visual representations; future display-rich environments and intelligent spaces.

The number of publications and websites related to visual representations has been steadily increasing, and a vast amount of literature and digital information currently exists on the subject. The cited references are selected for illustrating the ideas presented and are not necessarily the only significant contributions to the subject. This discussion is kept on a descriptive level; for a more thorough investigation students are encouraged to refer to the cited literature and websites.

Introduction and Rationale

In the early days of computer simulations, the output data from the simulation was presented in a table, or a matrix form, showing how the data was affected by various changes in the simulation parameters. With the emergence of graphics hardware and software tools, it was realized that visual representation, in the form of graphs, or even moving images or motion pictures generated from the data, is an effective way of interaction with the simulation [5, 6]. Attempts to enable portability and interoperability of visual presentations, and to speed up their generation, led to the development of **computer display standards** and **graphics file formats** [7, 8].

With the advent of high-capacity powerful computers, and the ever-increasing volumes of data generated by many simulations, the difficulties of managing, exploring, and analyzing simulation data has become a real challenge in many disciplines and applications. Visual representation enables a user-centered interactive discovery of new knowledge and alleviates the information overload. **Animations** can be used to experience a simulation in real-time (e.g., in training simulations).

The demand for more interactivity with visual simulations led to the use, and adaptation, of several technologies and facilities including virtual reality, virtual reality modeling language, and more recently serious games and virtual worlds. These are described briefly in subsequent sections, but their scope and applications go well beyond the scope of the present chapter. Today, visual representation of simulations has ever-expanding applications in the production, presentation and dissemination of simulation results in science, engineering, healthcare, business, education, art, sports, music, and other fields.

Rationale

There are several major forces driving the interest in, and the development of, visual representations [9]. These include:

- The human visual system can detect and discriminate between an incredibly diverse assortment of stimuli. This is because half the human brain is devoted, directly, or indirectly to vision (50% of the brain's neurons are associated with vision). Visualization engages the primary human sensory apparatus, vision, and extends it [10, 11]. Graphic output devices shift the burden of integrating information generated by computers onto the human vision system: the sensory channel with the highest capacity for distributed parallel processing. The old Chinese proverb "*A picture can express ten thousand words*" can now be extended to "*A picture can express ten thousand numbers or ten thousand gigabytes.*"

- The widespread availability of powerful graphics processors and high-performance computers.

- The ever-increasing size, dimensionality, and number of parameters of datasets being generated in numerical simulations, physical experiments, stock trading at stock exchanges, and medical imaging and acquisition devices (like MRI scanners), and the associated difficulty in understanding this data.

Visualization Pipeline

The process of creating visual representation from simulation data is described by the **visualization pipeline** concept. The pipeline describes a step-wise process involving four phases, namely:

- *Data Analysis*—Preparation of raw simulation data for visualization (e.g., by applying a smoothing filter or interpolating missing values). This step is computer centered, with little or no user interaction.

- *Filtering*—Selection of data portions to be visualized, this step is usually user-centered.

- *Mapping or Transformation*—Focus data are mapped to geometric primitives (e.g., points, lines) and their attributes (color, position, size). This is the most critical step for achieving effective visual representation.

- *Rendering*—Geometric data are transformed into visuals (e.g., pixel-based image in 2D, or a 3D model)

Much work has been devoted to the optimization of all steps of the visualization pipeline. This includes: semantic compression and feature extraction based on the raw data sets, adaptive visualization mappings that allow the user to choose between speed and accuracy, and exploiting new graphics hardware features for fast and high-quality rendering. Also, for very-large-scale simulations on multiprocessor computers, parallel adaptive rendering algorithms have been developed, along with parallel input-out (I/O) strategies to reduce the interframe delay.

In order to reduce redundant development work in visual representations, a number of generic and flexible visualization systems (extendable toolkits) have been developed, based on the visualization pipeline concept, and modifications thereof. Examples of these systems are the Visualization Toolkit (VTK), the Prefuse Visualization Toolkit (PVK), and the Insight Toolkit (ITK).

The VTK is an open source, freely available system supporting a variety of visualization algorithms, including scalar, vector, tensor, texture, and volumetric methods; advanced modeling techniques; several imaging algorithms that enable the user to mix 2D imaging/3D graphics, algorithms and data. Prefuse is a Java-based toolkit for building interactive information visualization applications. The Insight Toolkit focuses on medical applications. It covers a wide range of fundamental functions from image conversion and transformation, image segmentation, and analysis to geometric model generation and manipulation, all the way up to 3D visualization and interactive simulation. Several other flexible visualization systems have been developed for handling large datasets including, EnSight, ParaView, and AVS.

Categories of Visual Representations

A number of different classifications and taxonomies have been proposed for visual representations (VisRep). Some of these follow the classifications of simulations (e.g., stochastic or deterministic, steady-state or dynamic, continuous or discrete) [12]. The following four categories of visual representation will be examined:

- Scientific VisRep
- Data/Information VisRep
- Visual Analytics
- VisRep on Mobile Devices

Scientific VisRep

These are interactive visual displays of spatial/geometric data associated with scientific process. Since the early days of computer graphics, scientists and engineers used VisRep to investigate and explain physical phenomena and processes associated with their simulations. The publication of the 1987 NSF report on *Visualization in Scientific Computing* prompted researchers to investigate new approaches to the visualization process and also spawned the development of various visualization systems, as well as integrated software environments for visualization [13]. Some of the visualization systems addressed only specific application domains such as computational structures and fluid dynamics. However, a number of general systems evolved, some used a visual programming style interface.

In the mid-1980s graphics supercomputers and advanced visualization systems, incorporating modeling, animation, and rendering facilities, were introduced. Since the 1990s several Visual simulation-based modules have been developed as interactive multimedia tutorials for enhancing science and engineering education. To date, the fields of application of scientific VisRep include all engineering, natural, and medical sciences. A broad spectrum of numerical algorithms, techniques, and tools are currently available for scientific VisRep including contouring, isosurfaces, raycasting, volume rendering, and vector and tensor field visualization [14–16].

Grid computing (multiple, independent, networked computing clusters) has extended the horizons of computational simulations, by allowing aggregated computing resources to be harnessed in the solution of very large-scale problems. Visual representations are playing a crucial role in this activity by enabling the users to monitor and interpret the results in real-time, and to steer the computation (alter the values of the simulation parameter) as needed. Moreover, as the trend develops towards collaboration between geographically dispersed teams, computational steering will become a shared process, with participation by a range of teams at difficult locations.

Data/Information VisRep

This focuses on nongeometric data or information and involves selecting, transforming and representing the nongeometric data in a form that facilitates human interaction for exploration and understanding [17, 18]. Techniques have been developed to enable the user to modify the visual representation in real time, thus providing unparalleled perception of patterns and structural relations in the data.

The following major advantages have been identified for data/information VisRep:

- *Faster understanding*—VisRep supports perceptual inferences that are easy to recognize (e.g., pattern detection). Also, visual grouping of data reduces visual scanning and searching.
- *Increases cognitive capacity*—Perceptual inference off-loads cognitive processes, and visuals expand working memory.
- *Provides a big picture*—Large visual field enables view across more data.
- *Comprehends complexity*—VisRep enables exploration of complex relations and the creation of a cognitive model for the data.

Visual Analytics

This is an outgrowth of data/information and scientific VisRep and focuses on going beyond the interaction with the visual representation of the simulation to the analysis of data. The use of visual representations and interactions to accelerate rapid insight onto complex data is what distinguishes visual analytics from other types of analytical tools. Visual analytics tools incorporate technologies from many fields like knowledge management, statistical analysis, cognitive science, decision science, among others. The tools are used to maximize human capacity to perceive, understand, and reason about complex dynamic data and situations [19–21].

A definition of visual analytics was proposed in 2005 as *the science of analytical reasoning supported by the interactive visual interface.* It focuses on human-information interaction within massive, dynamically changing information spaces (which may be generated by simulations). It is an active research area concentrating on support of perceptual and cognitive operations that enable the user to *detect the expected and discover the unexpected* in complex information space. Although initial applications of visual analytics were in the homeland security and biology areas, current applications span several other fields.

Visual Representations on Mobile Devices

Wireless mobile devices are becoming ubiquitous due to their small size and their wide range of functionality. With their growing capabilities, they can provide useful visual representations of simulations in several applications including training, maintenance and repair, healthcare, emergency response, and management.

While the functionality of mobile devices is increasing, their limited processing, display, graphics, and power resources pose a major challenge in developing effective applications. Current research focuses on developing effective visual representations for mobile devices to enable the user to interact with the simulation and perform the task at hand with the least cognitive burden. Also, new methods are being developed to allow VisRep to scale between high-resolution large displays and small mobile display devices.

Virtual Reality and Virtual Worlds

Virtual reality and virtual worlds are examples of technologies that can be effectively used for visual representation of simulations, even though the motivations for their development, and their applications, go well beyond that.

Virtual reality (VR) is a technology that enables the user to interact with a 3D computer-simulated environment, be it real or an imagined one. Although the concept of VR may be traced back to the late 1950s, the term Virtual Reality was coined by Jaron Lanier in 1989. Early applications of VR allowed the user to have a relatively passive experience—watching a prerecorded animation while wearing a head-mounted display (HMD). Although users felt a sense of immersion (the feeling of being inside and part of the simulated environment), interactivity was limited to shifting the user's point of view by looking around [22–26].

Over the past 20 years, the level of sophistication of VR hardware, software, and systems has significantly increased and enabled the user to incorporate high levels of interaction and to use other human senses, in addition to sight, thereby enhancing the understanding of simulation data. With the advances in communication and collaboration technologies, geographically dispersed users can now interact, and jointly work together in a networked virtual environment [27]. Also, the concept of mixed reality was introduced as the merging of real and virtual worlds to generate new environment and visualization where physical and digital objects co-exist and interact in real time [28, 29].

Virtual Reality on the Web

In an attempt to enable real-time 3D interactive virtual reality experience on the web, Virtual Reality Modeling Language (VRML) was developed. Initially, it featured low-polygon, low-bandwidth rendering in an HTML environment. Subsequently, it became more sophisticated and evolved into 3D browsing and 3D gaming. Also, other languages and browsers were developed to enable the use of the web as a three-dimensional virtual space, including 3DML, X3D, and collaborative Design Activity (COLLADA).

Internet-based 3D Virtual Worlds

Virtual worlds are electronic environments that visually mimic complex physical spaces (emulation of physical world) and have game-like interfaces. Users are typically represented by animated characters and can interact with each other and with virtual objects in a cyberspace that uses the metaphor of the real world, but without its physical limitation [30]. Many current applications are exploring the potential of a 3D immersive virtual space from many angles (including, education, social networking, global commerce, and entertainment). They include World of Warcraft (a massively multiplayer online role-playing game—MMORPG), Second Life, The Croquet Project, and Google Earth. The latter builds a virtual structure, based on GIS and satellite imagery, on top of real earth data.

The fusion of virtualization and 3D web tools and objects in the environment is the new concept of **Metaverse space**, which is envisioned to propel the advances in several technologies, including simulation and visual representations [31].

Visual Representation Challenges/Current Focus

The major disciplines supporting visual representations, namely, scientific and information visualization, and visual analytics are relatively new disciplines. However, researchers in these fields have attempted to identify some of the challenges that are the focus of current research [17, 32, 33]. The challenges pertinent to VisRep include:

- *Creating effective, well-crafted visual representations* This is a labor-intensive process that requires a solid understanding of the visualization pipeline, characteristics of the data to be displayed, and the tasks to be performed. For scientific VisRep it includes developing better techniques and tools, particularly for multifield time-dependent, multiscale, and global-local simulations.

- *Facilitating knowledge discovery through information synthesis* This includes the integration of simulation data from heterogeneous multi-source, multi-dimensional, time-varying streams into a single seamless visual representation.

- *Incorporating uncertainty and error in the visual representation* Among the different ideas proposed are the use of uncertainty glyphs, isosurfaces with *error bars*, streamline with *standard deviations*, and volume visualizations with representations of confidence intervals.

- *Providing automated feature detection* For large-scale, complex simulations, automatically locating and representing important features of the simulation results is essential. Feature-detection involves both application-specific (e.g., finding and tracking vortices within CFD simulation) and application-independent aspects (e.g., comparing and tracking evolution of features in the simulation efficiently and robustly). Some of the commercial visual simulation packages have feature detection capabilities.

- *Effective use of novel hardware architecture and display technologies for VisRep*   These include advanced graphics processors, powerful collaborative platforms/access grid, high-resolution, and 3D novel displays. Also, collaborative visual representation algorithms, architectures, technologies, and environment are needed for decision making and support.

- *Developing new visual paradigms and visual representations based on cognitive and perceptual principles*   Augmenting the cognitive reasoning process with perceptual reasoning through VisRep permits the analytical reasoning process to become faster and more focused. It enables the thought process to translate from data to information, information to meaning, and meaning to understanding. Therefore, in designing effective visual representations, knowledge about the biophysics and psychophysical aspects of the visual system can be helpful.

A Look at the Future

Future work, learning, and research environments will be **ambient intelligent environments** in which humans are surrounded by computing and networking technology unobtrusively embedded in their surroundings (e.g., disappearing electronics, wireless sensors, and actuator networks, which are hidden in the background) [34, 35].

The environment will be capable of recognizing and responding to the presence of different individuals in a seamless, unobtrusive, and often invisible way. It will be adaptive with respect to a user's needs; capable of responding intelligently to spoken or gestured indications; and configurable in systems that are capable of engaging in intelligent dialogue.

Information technology will be diffused into everyday objects and settings, leading to completely new ways of supporting information workers in a variety of areas including individual productivity, team collaboration, and workflow.

Some of the key aspects of the future environment are a) the natural, intuitive, yet richly interactive interface for communication with the environment, b) the seamless integration of computers in the environment, c) the shared use of multiple displays by multiple users with multiple devices, and d) the use of computation to enhance the various activities in the environment.

Some of the display technologies are described subsequently, along with the characteristics of future smart rooms/intelligent spaces.

Future Display-Rich Environments

Future information-abundant display technologies will significantly enhance information portrayal, *enabling generation of insight at the speed of thought*. A spectrum of facilities will evolve ranging from small handheld devices and wearable displays to 3D autostereoscopic and volumetric displays to large high-resolution displays. The new display facilities will be integrated with sensors, computers, communication, and other hardware to form multi-display and multi-surface visual simulation and interaction environments. Among these display facilities are the following [36].

Autostereoscopic and Volumetric 3D Displays

Autostereoscopic displays provide 3D perception without the need of special glasses or other headgear. These displays are useful for applications in which the 3D depth perception is important (Figure 5-6a), such as scientific visualization of complex 3D systems.

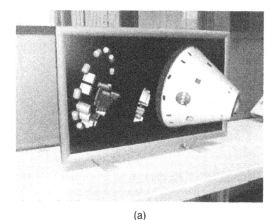

(a)

(b)

Figure 5-6   Autostereoscopic and volumetric displays: a) autostereoscopic display, b) volumetric display

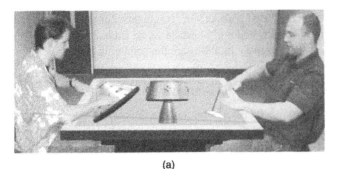

(a)

(b)

Figure 5-7   Tabletop displays: a) multiuser, multi-touch display, b) flatbed 3D display

Volumetric 3D displays are autostereoscopic displays that project volume filling three-dimensional imagery directly in space. Each volume element (or voxel) in a 3D scene emits visible light from the region in which it appears. Volumetric displays are useful for 360° viewing in fields like computer-aided design (CAD) (Figure 5-6b).

Tabletop Displays

These are designed for multi-user, small-group collaboration by providing a display interface that allows users to maintain eye contact while interacting with the display simultaneously (Figure 5-7a). Tabletops can be used as 3D autostereoscopic displays to provide a realistic touch and depth (Figure 5-7b).

Wearable Displays

These enable the users to visually interface with the digital world while navigating their physical environment. Among the various wearable display concepts are those a) integrating electronic fabric, ultra-thin flexible display with the user's clothes (Figure 5-8a) and b) those using a small LCD screen that hangs over one eye and fills the user's vision, providing the illusion of a larger (e.g., 10–20 inches) screen (Figure 5-8b).

Mobile Displays

With the mobile devices becoming increasingly powerful, visualization can make many mobile applications both more intuitive and productive. Also, the

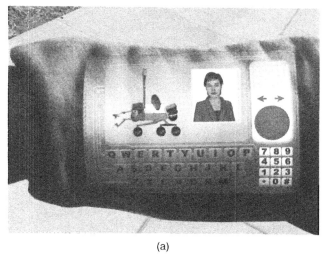

(a)

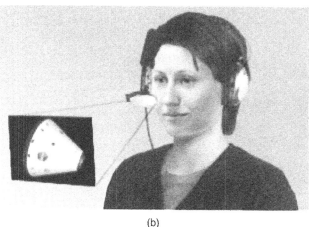

(b)

Figure 5-8   Wearable displays: a) ultra-thin flexible display on electronic fabric, b) tiny LCD screen that gives the illusion of a bigger screen

new software APIs, such as OpenGL ES, will enable more sophisticated mobile visualizations (Figure 5-9). However, these improvements will not eliminate most device limitations (such as the small screen size) or alter the mobility context.

Smart Rooms/Intelligent Spaces

These are interactive physical and digital spaces that have distributed networks of sensors, devices, processing elements and facilities that provide the users with

Figure 5-9 Mobile displays

Figure 5-10 Smart Room with multi-display and other facilities, including 3D autostereoscopic and holographic displays, tele-immersion, sketch interpretation, robotic computer, and wireless handheld computer

information and easy communication (Figure 5-10). The devices range from personal devices to public devices (e.g., intelligent interactive multi-display facilities, or thinking surfaces). The sensors and devices identify users, track ongoing activities, and recognize gestures and spoken commands in order to control the various displays and other facilities in the space.

The multi-display facilities provide the users with three modes of interaction and visualization, namely, independent, reflective, and coordinated views.

- *Independent view*   The displayed contents and the user inputs on the surfaces are not coupled or coordinated. Continuous input interactions on one surface are self-contained within that surface.

- *Reflective view*   The displayed contents and user inputs on multiple surfaces are tightly coupled. The surfaces share both visual content (pixels) and user interactions. Actions on one surface are directly reflected on other surfaces.

- *Coordinated multi-view*   The displayed contents and user inputs on multiple surfaces are interdependent, but not necessarily identical. Multiple surfaces share the same content, but the viewpoints are varied.

As the technology disappears, the devices in the space will no longer be perceived as computing elements, but rather as augmented elements of the physical environments. Personal devices are likely to be equipped with built-in multimodal interaction facilities (e.g., voice recognition and synthesis, pen-based pointing devices, and touch screen capability). In addition, users are able to navigate through virtual objects and interact with the environment, through human-like interfaces (including unencumbered, user-independent, deviceless voice, and vision multimodal input).

Human communication is not carried out on a single channel, but rather on many different channels that together convey many subtle messages and expressions. Multimodal interfaces provide the user with multiple models of interaction with the visual representation and enhance the usability: the weaknesses of one modality are offset by the strengths of another.

Also, combining 3D holographic displays and telepresence facilities can lead to new paradigms that significantly impact future science, engineering, business and other fields. An example of future application of cyber personalized medicine is depicted in Figure 5-11.

Dramatic improvements are on the horizon in visual representations and related technologies. The improvements are due in part to the developments in a number of leading-edge technologies and their synergistic couplings. The technologies include ubiquitous computing; ultra high-bandwidth networks; mobile/wireless communication; knowledge-based engineering; 3D and volumetric multiuser displays; large-scale media-rich, networked virtual environments; and other powerful visualization tools and facilities (e.g., high-resolution models and textures, soft subtle lighting, programmable pixel shaders, handheld visual simulation delivery systems, and collaborative visualization infrastructures).

The convergence of these technologies will enable the widespread availability of high-resolution, 3D visual representations with full immersion and interaction controls, as well as reshape work processes and collaboration between geographically dispersed teams. This will result in minimizing the time between generation of a hypothesis and the test of that idea, enable

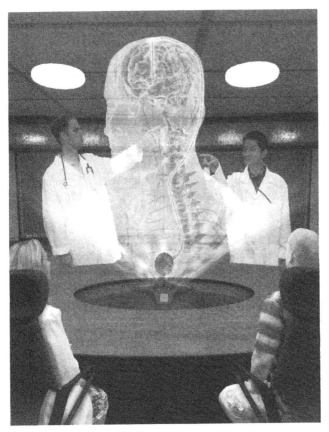

Figure 5-11 Cyber personalized medicine of the future. Application of telepresence and holographic displays permit real-time consultation (using an elaborate 3D visual simulation) with a patient, by geographically dispersed medical experts

insight at the speed of thought, and accelerate knowledge discovery and innovation.

CONCLUSION

Let's review the two M&S subtopics discussed in this chapter. *Simulation and Data Dependency* explored using simulation in support of analysis of a scientific investigation, which is the design and execution of a controlled experiment. Two examples, case studies, were provided to explain this concept: the security checkpoint at a school focusing on improving or alleviating the wait-time for cars to enter the campus and the minimum and maximum ticket prices for a football game. In both examples, simulation runs, statistical treatment of the outcomes, and the limits of the simulation (experiment) were expressed. We also learned that to support real experiments we must analyze the differences; we must build a confidence interval to measure performance. Both examples made us aware of simulation as a powerful tool for analysis because it can provide an inexhaustible source of data; however, that data must be carefully examined before proffering solutions in support of a hypothesis or problem.

The discussion on *Visual Representation* (VisRep) emphasized the importance of visualization, visual representations in the dissemination of simulation data. The subsection introduced the evolution of visual representations and the process of creating these representations known as the visualization pipeline. The narrative explained the four categories of VisRep: Scientific VisRep, Data/Information VisRep, Visual Analytics, and VisRep on Mobile Devices. We learned that virtual reality and virtual worlds are examples of technologies that can be effectively used for visual representation of simulations. The discussion listed six challenges facing visual representation including creating effective representations to integrating simulation data from differing sources to incorporating uncertainty and error into the representation. Future research and technical expertise will focus on ambient intelligent environments in which humans are surrounded by computing and networking technology unobtrusively embedded in their surroundings, future display-rich environments, and smart-rooms and intelligent spaces. Most assuredly, information technology will continue to be diffused into everyday objects and settings.

KEY TERMS

scientific method	experimental variables	Metaverse spaces
stochastic	confidence interval	ambient intelligence environments
Arena	computer display standards	autostereoscopic and volumetric 3D
randomness	graphics file formats	tabletop display
Lemma-Theorem Corollary structure	animations	wearable display
Central Limit theorem	grid computing	mobile display
measure of performance	virtual reality	
static variables	virtual worlds	

REFERENCES

1. MORSE P, KIMBALL G. *Methods of Operations Research*. Cambridge, MA: MIT Press; 1951.
2. KELTON D, SADOWSKI R, STURROCK D. *Simulation with Arena*. New York: McGraw-Hill; 2007.
3. LAW A, KELTON D. *Simulation Modeling and Analysis*. New York: McGraw-Hill; 2000.
4. BITINAS E, HENSHEID Z, TRUONG L. Pythagoras: A new agent-based simulation system. *Northrop Grumman Mission Systems Technology Review Journal* 2003;11(1):45–58.

5. CARLSON W. A Critical History of Computer Graphics and Animation. The Ohio State University, 2003; Available at http://design.osu.edu/carlson/history/lessons.html.

6. SHOAFF W. A Short History of Computer Graphics. August 30, 2000; Available at http://cs.fit.edu/~wds/classes/graphics/History/history/history.html.

7. BROWN CW, SHEPHERD BJ. *Graphics File Formats*. Greenwich, CT: Manning Publications Co; 1995.

8. Computer display standard. Available at http://en.wikipedia.org/wiki/Computer_display_standard. Accessed 2008 February 14.

9. van WIJK JJ (Dept. Mathematics and Computer Science, Technische Universiteit Eindhoven). The Value of Visualization. July 2005.

10. Visual perception. Available at http://en.wikipedia.org/wiki/Visual_perception. Accessed 2008 March 7.

11. DIAMANT E. *Modeling Visual Information Processing in the Brain: A Computer Vision Point of View and Approach*. Heidelberg: Springer Berlin; 2007.

12. TORY M, MÖLLER T (Graphics, Usability, and Visualization Lab; School of Computing Science, Simon Fraser University). Rethinking Visualization: A High-Level Taxonomy; 2004.

13. MCCORMICK BH, DEFANTI TA, BROWN MD. *Visualization in Scientific Computing*. New York: New York ACM Press; 1987.

14. BONNEAU G, ERTL T, NIELSON GM. *Scientific Visualization: The Visual Extraction of Knowledge from Data*. Boston, MA: Birkhäuser; 2006.

15. WRIGHT H. *Introduction to Scientific Visualization*. New York, NY: Springer; 2006.

16. HANSEN C, JOHNSON CR. *Visualization Handbook*. Burlington, MA: Academic Press; 2004.

17. SPENCE R. *Information Visualization: Design for Interaction*. Upper Saddle River, NJ: Prentice Hall; 2007.

18. CHEN C. *Information Visualization: Beyond the Horizon*. New York: Springer; 2006.

19. THOMAS J, COOK K, eds. *Illuminating the Path: The Research and Development Agenda for Visual Analytics*. Washington, DC: IEEE Computer Society; 2005.

20. FOLEY J, CARD S, EBERT D, MACEACHREN A, RIBARSKY B. Visual Analytics Education. IEEE Symposium On Visual Analytics Science and Technology, 2006 October 31–November 2.

21. KEIM DA, SCHNEIDEWIND J. *Introduction to the special issue on visual analytics*. Association for Computing Machinery. *Visual Analytics* ACM Greenwich, CT 2007;9(2):3–4.

22. VINCE J. *Introduction to Virtual Reality*. New York: Springer; 2004.

23. SHERMAN WR, CRAIG AB. *Understanding Virtual Reality: Interface, Application, and Design*. San Francisco, CA: Morgan Kaufmann; 2003.

24. GUTIERREZ M, VEXO F, THALMANN D. *Stepping into Virtual Reality*. New York: Springer; 2008.

25. BURDEA G, COIFFET P. *Virtual Reality Technology*. Washington, DC: Wiley-IEEE; 2003.

26. Immersive virtual reality. Available at http://en.wikipedia.org/wiki/Immersive_virtual_reality. Accessed 2007 October 20.

27. GREENHALGH C. *Large Scale Collaborative Virtual Environments*. New York: Springer; 1999.

28. BIMBER O, RASKAR R. *Spatial Augmented Reality: Merging Real and Virtual Worlds*. Wellesley, MA: A K Peters, Ltd.; 2005.

29. OHTA Y, HIDEYUKI TAMURA H, eds. *Mixed Reality: Merging Real and Virtual Worlds*. New York: Springer; 1999.

30. BARTLE R. *Designing Virtual Worlds*. NJ: Pearson Publ.; Upper Saddle River, 2003.

31. SMART J, CASCIO J, PAFFENDORF J, BRIDGES C, HUMMEL J, MOSS JHR. Metaverse Roadmap Overview. 2007. Available at http://www.metaverseroadmap.org/overview/.

32. RHYNE T, ed. Top Scientific Visualization Research Problems. Visualization Viewpoints, IEEE Computer Graphics and Applications, July/August 2004.

33. RHYNE T, et al. Can We Determine the Top Unresolved Problems of Visualization? Proceedings of the IEEE Visualization; 2004; IEEE Press; 2004. pp. 563–566.

34. WEBER W, RABAEY JM, AARTS EHL. *Ambient Intelligence*. New York: Springer; 2005.
35. RIVA G. *Ambient Intelligence: The Evolution of Technology, Communication and Cognition Towards the Future of Human-Computer Interaction*. Amsterdam, Netherlands: IOS Press; 2005.
36. NOOR AK. Ambient intelligence and its potential for future learning and virtual product creation environments. In: Topping BHV, Montero G, Montenegro R, eds. *Innovation in Engineering Computational Technology*. Glasgow: Saxe-Coburg Publications; 2006. pp. 1–22.

Chapter 6

Verification and Validation

Mikel D. Petty

INTRODUCTION

Verification and validation are important aspects of any simulation project; indeed, they are essential prerequisites to the credible and reliable use of a model and its results. But what are they exactly? What are the differences between them? What methods and procedures should be used to perform them? Those questions are taken up in this chapter.[1] This introductory section first motivates the need for verification and validation and provides definitions necessary to their examination. The next section discusses when and how to perform verification and validation. Following that, a set of examples of verification and validation are presented. A summary and conclusions ends the chapter.

Motivation

In the civil aviation industry in the United States and in other nations, a commercial airline pilot may be qualified to fly a new type of aircraft after training to fly that aircraft type solely in flight simulators (the simulators must meet official standards) [1, 2]. Thus it is entirely possible that the first time a pilot actually flies an aircraft of the type for which he or she has been newly qualified there will be passengers on board; people who, quite understandably, have a keen personal interest in the qualifications of that pilot. Conversion to a new aircraft type after simulator training only is based on an assumption that seems rather bold: the flight simulator in which the training took place is sufficiently accurate with respect to its recreation of the

---

[1] This chapter is an introductory tutorial on verification and validation. For a more comprehensive and technically oriented survey of the same topic, see [6]. For even more detail, a very large amount of information on verification and validation, including concept documents, method taxonomies, glossaries, and management guidelines has been assembled by the U.S. Department of Defense [26].

Principles of Modeling and Simulation: A Multidisciplinary Approach, Edited by John A. Sokolowski and Catherine M. Banks.
Copyright © 2009 John Wiley & Sons, Inc.

flight dynamics, performance, and controls of the aircraft type in question that prior practice in an actual aircraft is not necessary.

As in this example, simulations are often taken to be accurate to a degree that justifies a rather large risk, whether personal or financial. Clearly, the assumption of accuracy is not made purely on the basis of the good intentions of the developers. In properly conducted simulation projects the accuracy of the simulation, and the models upon which the simulation is based, is assessed, measured, and established via verification and validation. Verification and validation are processes, performed using methods suited to the model and to an extent appropriate for the application that measure the accuracy of models.

Background Definitions

Several background definitions are needed to support an effective explanation of verification and validation. These definitions are based on the assumption that there is some real-world system, such as an aircraft, which is to be simulated for some known application, such as flight training.

Simuland

A **simuland** is the real-world system of interest. It is the object, process, or phenomenon to be simulated. The simuland might be the aircraft in a flight simulator, the assembly of automobiles in a factory assembly line simulation, or ocean water evaporation in a climate simulation. The simuland may be understood to include not only the specific object of interest, but also any other aspects of the real-world that affect the object of interest in a significant way. For example, for a flight simulator the simuland could include not just the aircraft itself but weather phenomena that affect the aircraft's flight. Simulands need not actually exist in the real world; for example, in combat simulation hypothetical nonexistent weapons systems are often modeled to analyze how a postulated capability would affect battlefield outcomes.[2]

Model

In general terms, a **model** is a representation of something else, e.g., a fashion model representing how a garment might look on a prospective customer. In modeling and simulation, a model is a representation of a simuland.[3] Models are often developed

---

[2] In such applications, the term of practice used to describe a hypothetical nonexistent simulands is *notional*.

[3] A specialized term omitted from the main list of definitions is *referent*. A referent is the body of knowledge that the modelers have about the simuland. The referent may include everything from quantitative formal knowledge, such as engineering equations describing an aircraft engine's thrust at various throttle settings, to qualitative informal knowledge, such as an experienced pilot's intuitive expectation for the feeling of buffet that occurs just before a high-speed stall. Because the referent is by definition everything the modeler knows about the simuland, a model is arguably a representation of the referent, not the simuland. However, this admittedly pedantic distinction is not particularly important here, and model accuracy will be discussed with respect to the simuland, not the referent.

with their intended application in mind, thereby emphasizing characteristics of the simuland considered important for the application and deemphasizing or omitting others. Models may be in many forms, and the modeling process often involves developing several different representations of the same simuland or of different aspects of the same simuland. Here, models will be broadly grouped into two types: *conceptual* and *executable*. **Conceptual models** document those aspects of the simuland that are to be represented and those that are to be omitted.[4] Information contained in a conceptual model may include the physics of the simuland, the objects and environmental phenomena to be modeled, and representative use cases. Several different forms or documentation, or combinations of them, may be used for conceptual models, including mathematical equations, flowcharts, UML (Unified Modeling Language) diagrams, data tables, or expository text [3]. The **executable model**, as one might expect, is a model that can be executed. The primary example of an executable model considered here is a computer program. Execution of the executable model is intended to simulate the simuland as detailed in the conceptual model, so the conceptual model is thereby a design specification for the executable model.

Simulation

Simulation is the process of executing a model (an executable model, obviously) over time. Here time may mean simulated time for those models that model the passage of time, such as a real-time flight simulator, or execution time for those models that do not model the passage of time, such as a Monte Carlo simulation [4]. The process of running a flight simulator is simulation. The term may also refer to a single execution of a model, as in "During the last simulation, the pilot was able to land the aircraft successfully."[5]

Results

The results are the output produced by a model during a simulation. The results may be available to the user during the simulation, such as the out-the-window views generated in real-time by a flight simulator, or at the end of the simulation, such as the queue length and wait time statistics produced by a discrete-event simulations described in Chapters 3–5. Regardless of when they are available and what form they take, a model's results are very important as the primary object of validation.

---

[4] The term *conceptual model* is used in different ways in the literature. Some define a conceptual model as a specific type of diagram (e.g., UML class diagram) or documentation of a particular aspect of the simuland (e.g., the classes of objects in the environment of the simuland and their interactions), whereas others define it more broadly to encompass any non-executable documentation of the aspects of the simuland to be modeled. It is used here in the broad sense, a use of the term also found in [9].

[5] The term *simulation* is also often used in a third sense to refer to a large model, perhaps containing multiple models as subcomponents or submodels. For example, a large constructive battlefield model composed of various submodels, including vehicle dynamics, intervisibility, and direct fire, might be referred to as a simulation. In this chapter this sense of simulation is avoided, with the term model used regardless of the size and number of submodels.

Requirements

When developing a model of a simuland, it is typically not necessary to represent all aspects of the simuland in the model, and for those that are represented, it is typically not necessary to represent all at the same level of detail and degree of accuracy.[6] For example, in a combat flight simulator with computer-controlled hostile aircraft, it is generally not required to simulate whether the enemy pilots are hungry or not, and while ground vehicles may be present in such a simulator (perhaps to serve as targets) their driving movement across the ground surface will likely be modeled with less detail and accuracy than the flight dynamics of the aircraft. With respect to verification and validation, the requirements specify which aspects of the simuland must be modeled, and for those to be included, how accurate the model must be. The requirements are driven by the intended application.

To illustrate these definitions and to provide a simple example which will be returned to later, a simple model is introduced. In this example, the simuland is a phenomenon, namely gravity. The model is intended to represent the height over time of an object freely falling to the earth. A mathematical (and nonexecutable) model is given by the following equation:

$$h(t) = -4.9t^2 + vt + s$$

where

t = time elapsed since the initial moment, when the object began falling (seconds)

v = initial velocity of the falling object (meters/second), with positive values indicating upward velocity

s = initial height of the object (meters)

-4.9 = change in height of the object due to gravity (meters), with the negative value indicating downward movement

$h(t)$ = height of the object at time t (meters).

This simple model is clearly not fully accurate, even for the relatively straight-forward simuland it is intended to represent. Both the slowing effect of air resistance and the reduction of the force of gravity at greater distances from the earth are absent from the model. It also omits the surface of the earth, so applying the model with a value of t greater than the time required for the object hit the ground will give a nonsensical result.

An executable version of this model of gravity, given as C code follows. This code is assuredly not intended as an example of good software engineering practices, as v and s are hardcoded in the model. Note that the code does consider the surface of the earth, stopping the model once a nonpositive height is reached, so in

---

[6] The omission or reduction of detail not considered necessary in a model is referred to as *abstraction*. The term *fidelity* is also often used to refer to a model's accuracy with respect to the represented simuland. Here accuracy and fidelity are used interchangeably.

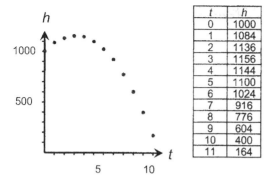

t	h
0	1000
1	1084
2	1136
3	1156
4	1144
5	1100
6	1024
7	916
8	776
9	604
10	400
11	164

Figure 6-1   Results of the simple gravity model

that particular way it has slightly more accuracy than the earlier mathematical model.

```
/* Height of a falling object. */
main()
{
 float h, v = 100.0, s = 1000.0;
 int t;
 for (t = 0, h = s; h >= 0.0; t++)
 {
  h = (-4.9 * t * t) + (v * t) + s;
  printf("Height at time %d = %f\n", t, h);
 }
}
```

The process of executing this model is simulation. The results produced by the simulation are shown in both graphical and tabular form in Figure 6-1. The results assume the initial velocity and height hard coded into the program, e.g., the object was initially propelled upward at 100 meters/second from a vantage point 1000 meters above the ground. After that initial impetus, the object falls freely. The parabolic curve in the figure is not meant to suggest that the object is following a curved path; in this simple model, the object may travel only straight up and straight down. The horizontal axis in the graph in the figure is time, and the curve shows how the height changes over time.

Verification and Validation Definitions

Of central concern in this chapter are *verification* and *validation*. These terms have meanings in a general quality management context as well as in the specific modeling and simulation context, and in both cases, the latter meaning can be understood a modeling and simulation special case of the more general meaning. These definitions, as well as that of the related term *accreditation*, follow.

Verification

In general quality management, **verification** refers to a testing process that determines whether a product is consistent with its specifications or compliant with applicable regulations. In modeling and simulation, verification is typically defined analogously, as the process of determining if an implemented model is consistent with its specification [5]. Verification is also concerned with whether the model as designed will satisfy the requirements of the intended application. Verification is concerned with transformational accuracy, i.e., that of transforming the model's requirements into a conceptual model and the conceptual model into an executable model. The verification process is frequently quite similar to that employed in general software engineering, with the modeling aspects of the software entering verification by virtue of their inclusion in the model's design specification. Typical questions to be answered during verification include:

1. Does the program code of the executable model correctly implement the conceptual model?
2. Does the conceptual model satisfy the intended uses of the model?
3. Does the executable model produce results when it is needed and in the required format(s)?

Validation

In general quality management, **validation** refers to a testing process that determines whether a product satisfies the requirements of its intended customer or user. In modeling and simulation, validation is the process of determining the degree to which the model is an accurate representation of the simuland [5]. Validation is concerned with representational accuracy, i.e., that of representing the simuland in the conceptual model and the results produced by the executable model. The process of validation assesses the accuracy of the models. The accuracy needed should be considered with respect to its intended uses, and differing degrees of required accuracy may be reflected in the methods used for validation. Typical questions to be answered during validation include:

1. Is the conceptual model a correct representation of the simuland?
2. How close are the results produced by the executable model to the behavior of the simuland?
3. Under what range of inputs are the model's results credible and useful?

Accreditation

Accreditation, although often grouped with verification and validation in the modeling and simulation context in the common phrase "verification, validation, and accreditation" (or "VV&A"), is an entirely different sort of process from the others. Verification and validation are fundamentally testing processes, and are technical in nature. Accreditation, on the other hand, is a decision process, and is nontechnical in nature, though it may be informed by technical data. Accreditation is the official certification by a responsible authority that a model is acceptable for use for a specific purpose [5]. Accreditation is concerned with official usability, i.e., the determination that the model may be used. Accreditation is always for a specific purpose, such as a particular training exercise or analysis experiment, or a particular class of applications. Models should not be accredited for "any purpose," because an overly broad accreditation could result in a use of a model for an application for which it has not been validated or is not suited. The accrediting authority typically makes the accreditation decision based on the findings of the verification and validation processes. Typical questions to be answered during accreditation include:

1. Are the capabilities of the model and requirements of the planned application consistent?

2. Do the verification and validation results show that the model will produce usefully accurate results if used for the planned application?

3. What are the consequences if an insufficiently accurate model is used for the planned application?

To summarize these definitions, note that verification and validation are both testing processes, but they have different purposes.[7] The difference between them is often summarized with these questions: verification asks "Was the model made right?" whereas validation asks "Was the right model made?" [6, 7]. Continuing this theme, accreditation asks "Is the model the right one for the job?"

Verification and Validation as Comparisons

As discussed in the definitions, verification and validation are both ultimately activities that compare one thing to another; the defining difference between them is what things are being compared. Figure 6-2 illustrates these ideas. In the figure, the boxes represent the objects or artifacts involved in a simulation project.[8] The solid arrows connecting them represent processes that produce one object or artifact by transforming or using another. The results, for example, are produced by executing the executable model. The dashed arrows represent comparisons between the artifacts.

[7] Verification and validation are concerned with accuracy (transformational and representational, respectively), which is only one of several aspects of quality in a simulation project; others include execution efficiency, maintainability, portability, reusability, and usability (user friendliness) [6].

[8] Everything in the boxes in Figure 6-2 is an *artifact* in the sense used in [27], i.e., an intermediate or final product produced during the project, except the simuland itself, hence the phrase *objects or artifacts*. Hereinafter the term *artifacts* may be used alone with the understanding that it includes the simuland. Of course the simuland could itself be an artifact of some earlier project, e.g., an aircraft is an artifact, but in the context of the simulation project it is not a product of that project.

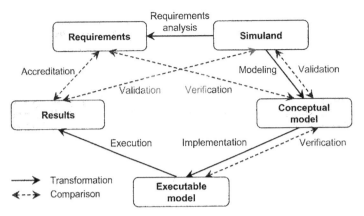

Figure 6-2   Comparisons in verification, validation, and accreditation

Verification refers to either of two comparisons to the conceptual model. The first comparison is between the requirements and the conceptual model. In this comparison, verification seeks to determine if the requirements of the intended application will be met by the model described in the conceptual model. The second comparison is between the conceptual model and the executable model. The goal of verification in this comparison is to determine if the executable model, typically implemented as software, is consistent and complete with respect to the conceptual model.

Validation likewise refers to either of two comparisons to the simuland. The first comparison is between the simuland and the conceptual model. In this comparison, validation seeks to determine if the simuland, and in particular those aspects of the simuland to be modeled, have been accurately and completely described in the conceptual model. The second comparison is between the simuland and the results.[9] The goal of validation in this comparison is to determine if the results, which are the output of a simulation using the executable model, are sufficiently accurate with respect to the actual behavior of the simuland, as defined by data documenting its behavior. Thus validation compares the output of the executable model with observations of the simuland.

PERFORMING VERIFICATION AND VALIDATION

Having defined verification and validation and understood them as both testing processes and comparisons between objects or artifacts in a simulation project, there remains the important question of how verification and validation are actually done. In verification, general software engineering methods that compare the implemented

---

[9] To be precise, the results are not compared to the simuland itself, but to observations of the simuland. For example, it is not practical to compare the height values produced by the gravity model to the height of an object as it falls; rather, the results are compared to data documenting measured times and heights recorded while observing the simuland.

model with its design specification are often applicable. In contrast, for validation specialized methods that compare the results of executing the model with the simuland are frequently needed. The following sections provide a discussion of verification and validation methodologies, beginning with a discussion of when to use them.

Verification and Validation within a Simulation Project

Verification and validation do not take place in isolation; rather, they are done as part of a simulation project. To place verification and validation in the context of a project and answer the question of when to do them, it is common in the research literature to find specific verification and validation activities assigned specific phases of a project (e.g., [7]).[10] However, these recommendations are rarely consistent in detail, for two reasons. First, there are different types of simulation projects (simulation studies, simulation software development, and simulation events are examples) which have different phases and different verification and validation processes. Even when considering a single type of simulation project, there are different project phase breakdowns and artifact lists in the literature. For examples, compare the project phases for simulation studies given in [5, 6, 8, 9], all of which make sense in the context of the individual source. Consequently, the association of verification and verification activities with the different phases and comparison with the different artifacts inevitably produces different recommended processes.[11]

Nevertheless, it is possible to generalize about the different project phase breakdowns. Figure 6-2 can be understood as a simplified example of such a breakdown, as it suggests a sequence of activities (the transformations) that produce intermediate and final products (the artifacts) over the course of the simulation project. The more detailed phase breakdowns cited earlier typically define more project phases, more objects and artifacts, and more verification and validation comparisons between the artifacts than those shown in Figure 6-2.[12] Going further into an explanation and

---

[10] Simulation project phase breakdowns are often called *simulation life cycles*, e.g., [6].

[11] Further complicating the issue is that there are two motivations for guidelines about performing verification and validation found in the literature. The first motivation is technical effectiveness; technically motivated discussions are concerned with when to perform verification and validation activities and which methods to use based on artifact availability and methodological characteristics, with an ultimate goal of ensuring model accuracy. The second motivation is official policies; policy-motivated discussions relate official guidelines for when to perform verification and validation activities and how much effort to apply to them based on organizational policies and procedures, with an ultimate objective of receiving model accreditation. Of course, ideally the two motivations are closely coupled, but the distinction should be kept in mind when reviewing verification and validation guidelines. In this chapter, only technical effectiveness is considered.

[12] For example, [6] identifies the *conceptual model* and the *communicative model*, both of which are forms or components of the *conceptual model* in Figure 6-2, and includes a verification comparison between them. Similarly, [6] also identifies the *programmed model* and the *experimental model*, both of which are forms or components of the *executable model* in Figure 6-2, and includes another verification comparison between them.

reconciliation of the various simulation project types and project phase breakdowns detailed enough to locate specific verification and validation activities within them is beyond the scope of this chapter. In any case, although the details and level of granularity differ, the overall concept and sequence of the breakdowns are essentially the same.

Despite the detailed differences between the various available simulation project phase breakdowns found in the literature, and the recommendations for when to perform verification and validation within the project, there is a underlying guideline that is common across all of them: verification and validation in general, and specific verification and validation comparisons in particular, should be conducted as soon as possible. (Two of the 15 "Principles of VV&A" in get at this idea [6].) But when is "as soon as possible"? The answer is as soon as the artifacts to be compared in the specific comparison are available. For some verification activities, both the conceptual model (in the form of design documents) and the executable model (in the form of programming language source code) are needed. For some validation activities, the results of the simulation, perhaps including event logs and numerical output, as well as the data representing observations of the simuland, are required. The detailed breakdowns of verification and validation activity by project phase found in the literature are consistent with the idea that verification and validation should proceed as soon as the artifacts to be compared are available, assigning to the different project phases specific verification and validation activities and methods (methods will be discussed later) that are appropriate to the artifacts available in that phase.

Verification and Validation Methods

Having determined when to do verification and validation, the simulationist must also decide how. There are a surprisingly large variety of verification and validation methods. The diversity is due to the range of different simulation project types, different artifacts produced during the course of simulation projects, different subjects (simulands) of those projects, and different types of data available for those subjects. Some of the methods (especially verification methods) come from software engineering, because the executable models in simulation projects are almost always realized as software, while others (especially validation methods) are specific to modeling and simulation, and typically involve data describing the simuland. Over 90 different verification and validation methods, grouped into four categories (informal, static, dynamic, and formal), are listed and individually described in (and that list, while comprehensive, is not complete); repeating each of the individual method descriptions here would be pointlessly duplicative [6]. Instead, the four categories from that source will be defined and representative examples from each category described. In the next section, additional examples of verification and validation methods will be given as case studies.

Informal

Informal verification and validation methods are more qualitative than quantitative, and generally rely heavily on subjective human evaluation, rather than detailed mathematical analysis. Experts will examine some artifact of the simulation project, e.g., a model expressed as UML diagrams, or the simulation results, e.g., variation in service time in a manufacturing simulation, and assess the model based on that examination and their reasoning and expertise. Two examples of informal methods are *inspection* and *face validation.*

Inspection is a verification method that compares project artifacts. In inspection, organized teams of developers and testers inspect model artifacts, such as design documents, algorithms, physics equations, and programming language code. Based on their own expertise the inspectors manually compare the artifacts being inspected with the appropriate object of comparison, e.g., programming language code (the executable model) might be compared with algorithms and equations (the conceptual model). The persons doing the inspection may or may not be the developers of the model being inspected, depending on the resources of the project and the developing organization. Inspections may be highly structured, with members of an inspection team assigned specific roles, such as moderator, reader, and recorder, and specific procedure steps used in the inspection [6]. The inspectors identify, assess, and prioritize potential faults in the model.

Face validation is a validation method that compares simuland behavior to model results. In face validation, observers who may be potential users or subject matter experts with respect to the simuland review or observe the results of a simulation (an execution of the executable model). Based on their knowledge of the simuland, the observers subjectively compare the behavior of the simuland as reflected in the simulation results with their knowledge of the behavior of the actual simuland under the same conditions, and judge whether the former is acceptably accurate. Face validation is frequently used in real-time virtual simulations where they experience of a user interacting with the simulation is an important part of its application. An example of this form of face validation is an evaluation of the accuracy of a flight simulator's response to control inputs by having an experienced pilot fly the simulator through a range of maneuvers. Face validation is often used as a validation method of last resort, when a shortage of time or a lack of reliable data describing simuland behavior precludes the use of more objective and quantitative methods. While moving beyond face validation to more objective and quantitative methods should always be a goal, face validation is clearly preferable to no validation.

Static

Static verification and validation methods involve assessment of the model's accuracy on the basis of characteristics of the model and executable model that can be determined without execution of a simulation. Static techniques often involve analysis of the programming language code of the implemented model, and may be supported by automated tools to perform the analysis or manual notations or diagrams to support it. Static methods are more often performed by developers and other technical experts, as compared to informal methods, which depend more on subject matter experts. Two examples of static methods are *data analysis* and *cause-effect graphing*.

Data analysis is a verification method that compares data definitions and operations in the conceptual model to those in the executable model. Data analysis ensures that data is properly defined (correct data types, suitable allowable data ranges) and that proper operations are applied to the data structures in the executable model. Data analysis includes data dependency analysis (analyzing which data variables depend on which other variables) and data flow analysis (analyzing which variables are passed between modules in the executable model code).

Cause-effect graphing is a validation method that compares cause-and-effect relationships in the simuland to those in the conceptual model. Causes are events or conditions that may occur in the simuland, and effects are the consequences or state changes that result from the causes. For example, lowering flaps in a flight simulator (a cause) will change the flight dynamics of the aircraft, increasing both drag and lift (the effects). Note that effects may themselves be causes of further effects; e.g., the additional drag caused by lowering flaps will then cause a slowing of the aircraft. In cause-effect graphing, all causes and effects considered to be important in the intended application of the model are identified in the simuland and in the conceptual model and compared; missing and extraneous cause-effect relationships are corrected. Causes and effects are documented and analyzed through the use of cause-effect graphs, which are essentially directed graphs where causes and effects, represented by nodes in the graph, are connected by the effects that related them, represented by edges.

Dynamic

Dynamic verification and validation methods assess model accuracy by executing the executable model and evaluating the results. The evaluation may involve comparing the results with data describing the behavior or the simuland or the results of other models. Because the comparisons in dynamic methods are typically of numerical results and data, dynamic methods are generally objective and quantitative. Statistical validation methods, of which nearly 20 are identified in [6], are in this category. Two examples of dynamic methods are *sensitivity analysis* and *predictive validation*.

Sensitivity analysis is a validation method that compares magnitude and variability in simuland behavior to magnitude and variability in the model results. It is an analysis of the range and variability in model results. A test execution of the model is arranged so as to cause the inputs to the model to vary over their full allowable range. The magnitude and variability of the results produced are measured and compared to the magnitude and variability of the simuland's behavior over the same range of input values. Differences could suggest invalidity in the model; if there are significant differences for some input values but not for others, this could suggest invalidity for some specific ranges of inputs. If sufficient data regarding the simuland is available, sensitivity analysis can be conducted by comparing the response surfaces of the model and the simuland for appropriately chosen independent variables (the input values) and dependent variables (the output results); the sign and magnitude of the difference between the two response surfaces can be calculated and analyzed [10]. Beyond validation, sensitivity analysis can also be used to evaluate model response to errors in the input and to establish which inputs have the greatest impact on the results, information which can focus efforts and establish accuracy requirements when preparing model input data [11].

Predictive validation is a validation method that compares specific outcomes in simuland behavior to corresponding outcomes in the model results. Predictive validation may be used when available information about the behavior of the simuland includes corresponding input and output values, i.e., historical or experimental data is available that shows how the simuland behaved under well established conditions. Given such data, the model is executed with the same inputs and its results are compared with the historical or experimental data.[13] For example, the airspeed of an aircraft at different altitudes and throttle settings in a flight simulator can be compared to actual flight test for the aircraft being modeled, if the latter is available. Similarity or dissimilarity between the simuland's behavior and the simulation's results suggest validity or invalidity. The actual comparison may be done in a variety of ways, some of which are considered validation methods in their own right (e.g., statistical methods); this is certainly acceptable and strengthens the validation power of the method. Because it is based on a direct comparison between simuland behavior and model results, predictive validation is a valuable method.

Formal

Formal verification and validation methods employ mathematical proofs of correctness to establish model characteristics. Statements about the model are developed using a formal language or notation and manipulated using logical rules; conclusions derived about the model are unassailable from a mathematical perspective. However, formal methods are quite difficult to apply in practice, as the complexity of most useful models is too great for current tools and methods to deal with practically [6]. Nevertheless, useful results can be achieved using formal methods in some constrained situations. Two examples of formal methods are *inductive assertions* and *predicate calculus*.

Inductive assertion is a verification method that compares the programming language code for the executable model to descriptions of the simuland in the

---

[13] The method is called predictive validation because the model is executed to "predict" the simuland's behavior. However, because the events being predicted by the model are in the past, some prefer to call the process *retrodiction* and the method *retrodictive validation*.

conceptual model. It is closely related to techniques from program proving. Assertion statements, which are statements about the input-to-output relations for model variables that must be true for the executable model to be correct, are associated with the beginning and end of each of the possible execution paths in the executable model. If it can then be proven for each execution path that the truth of beginning assertion and the execution of the instructions along the path imply the truth of the ending assertion, then the model is proven correct. The proofs are done using mathematical induction.

Predicate calculus is a validation method that compares the simuland to the conceptual model.[14] Predicate calculus is a formal logic system that allows creation, manipulation, and proof of formal statements that describe the existence and properties of objects. Characteristics of both the simuland and the conceptual model can be described using predicate calculus. One procedure of predicate calculus is the proving of arguments, which can demonstrate that one set of properties of the object in question, if true, together imply additional properties. The goal of the method is that by describing properties of the simuland and conceptual model using predicate calculus, it can be possible to prove that the two are consistent.

Validation Risks, Bounds of Validity, and Model Credibility

What types of errors may occur during verification and validation, and what risks follow from those errors? Figure 6-3 summarizes the types of verification and validation errors and risks.[15]

	Model valid	Model not valid	Model not relevant
Results accepted, model used	Correct	**Type II error** Use of invalid model; Incorrect V&V; Model user's risk; **More** serious error	**Type III error** Use of irrelevant model; Accreditation mistake; Accreditor's risk; **More** serious error
Results not accepted, model not used	**Type I error** Non-use of valid model; Insufficient V&V; Model builder's risk; **Less** serious error	Correct	Correct

Figure 6-3   Verification and validation errors

[14] Predicate calculus is also known as first order predicate calculus and predicate logic.

[15] Figure 6-3 is adapted from a flowchart that shows how the errors might arise found in [6].

In the figure, three possibilities regarding the model's accuracy are considered; it may be accurate enough to be used for the intended application ("valid"), it may not be accurate enough ("not valid") or it may be not relevant to the intended application. Two possibilities regarding the model's use are considered; the model's results may be accepted and used for the intended application, or they may not. The correct decisions are, of course, when a valid model is used or when an invalid or irrelevant model is not used.

A Type I error occurs when a valid model is not used. For example, a valid flight simulator is not used to train and qualify a pilot. This may be due to insufficient validation to persuade the accrediting authority to certify the model. A Type I error can result in model development costs that are entirely wasted if the model is never used or needlessly increased if model development continues [6]. Additionally, whatever potential benefits that using the model might have conferred, such as reduced training costs or improved decision analyses, are forfeited. The likelihood of a Type I error is termed *model builder's risk* [12].

A Type II error occurs when an invalid model is used. For example, an invalid flight simulator is used to train and qualify a pilot. This may occur when validation is done incorrectly but convincingly, erroneously persuading the accrediting authority to certify the model for use. A Type II error can result in disastrous consequences, such as an aircraft crash because of an improperly trained pilot or a bridge collapsing because of faulty analyses of structural loads and stresses. The likelihood of a Type II error is termed *model user's risk* [12].

A Type III error occurs when an irrelevant model, i.e., one not appropriate for the intended application, is used. This differs from a Type II error, where the model is relevant but invalid; in a Type III error the model is in fact valid for some purpose, but it is not suitable for the intended application. For example, a pilot may be trained and qualified for an aircraft type in a flight simulator valid for some other type. Type III errors are distressingly common; models that are successfully used for their original applications can acquire an unjustified reputation for broad validity, tempting project managers eager to reduce costs by leveraging past investments to use the models inappropriately. Unfortunately, the potential consequences of a Type III error are similar, and thus similarly serious, to those of a Type II error. The likelihood of a Type III error is termed *model accreditor's risk*.

Reducing validation risk can be accomplished, in part, by establishing a model's bounds of validity. The goal of verification and validation is not to simply declare "the model is valid," because for all but the simplest models such a simple and broad declaration is inappropriate. Rather, the goal is to determine when (i.e., for what inputs) the model is usefully accurate, and when it is not, a notion sometimes referred to as the model's bounds of validity. The notion of bounds of validity will be illustrated using the example gravity model. Consider these three versions of a gravity model:

1. $h(t) = 776$
2. $h(t) = (-420/9)t + 1864$
3. $h(t) = -4.9t^2 + vt + s$

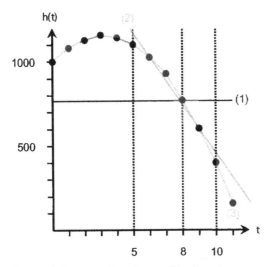

Figure 6-4 Results from three models of gravity

In these models, let $v = 100$ and $s = 1000$.

The results (i.e., the heights) produced by these models for time values from 0 to 11 are shown in Figure 6-4. Model 1 is an extremely simple and low fidelity linear model; it always returns the same height regardless of time. It corresponds to the horizontal line in the figure. Essentially by coincidence, it is accurate for one time value ($t = 8$). Model 2 is a slightly better linear model, corresponding to the downward sloping line in Figure 6-4. As can be seen in the figure, Model 2 returns height values that are reasonably close to correct over a range of time values (from $t = 5$ to $t = 10$). Model 3 is the original example gravity model, which is quite accurate within its assumptions of negligible air resistance and proximity to the surface of the earth.

Assume that the accuracy of each of these three models was being determined by a validation process that compared the models' results with observations of the simuland, i.e., measurements of the height of objects moving under gravity. If the observations and validation were performed only for a single time value, namely $t = 8$, Model 1 would appear to be accurate. If the observations were performed within the right range of values, namely $5 \le t \le 10$, then the results of Model 2 will match the observations of the simuland fairly well. Only validation over a sufficient range of time values, namely $0 \le t \le 11$, would reveal Model 3 as the most accurate.

These three models are all rather simple, with only a small range of possible inputs, and the accuracy of each is already known. Performing validation in a way that would suggest that either Model 1 or Model 2 was accurate might seem to be unlikely. But suppose the models were 1000 times more complex, a level of complexity more typical of practical models, with a commensurately expanded range of input values. Under these conditions, it is more plausible that a validation effort constrained by limited resources or data availability could consider Models 1 or 2 to be accurate.

The conclusion to be drawn from this example has several related points. First, validation should be done over the full range of input values expected in the intended use. Only by doing so would the superior accuracy of Model 3 be distinguished from Model 2 in the example. Second, one objective of validation is to determine the range of inputs over which the model is accurate enough to use, i.e., to determine the bounds of validity. In the example, there were three models to choose from. More often, there is only one model and the question is thus not which model is most accurate, but rather when (that is, for what inputs) the one available model is accurate enough. In this example, if Model 2 or its counterpart that is 1000 times more complex, is the only one available, an outcome of the validation process would be a statement that it is accurate within a certain range of time values. Finally, finding during validation that a model is accurate only within a certain range of inputs is not necessarily a disqualification of the model; it is entirely possible that the intended use of that model will only produce inputs within that range, thus making the model acceptable. The validity of a model depends on its application. However, even when the range of acceptable inputs found during validation is within the intended use, that range should be documented. Otherwise, later reuse of the model with input values outside the bounds of validity could unknowingly produce inaccurate results.

Model credibility can be understood as a measure of how likely a model's results are to be considered acceptable for an application. Verification, validation, and accreditation all relate to credibility; verification and validation are processes that contribute to model credibility, and accreditation is an official recognition that a model has sufficient credibility to be used for a specific purpose. Developing model credibility requires an investment of resources in model development, verification, and validation; in other words, credibility comes at a cost. Figure 6-5 suggests the relationship between model cost, credibility, and utility.

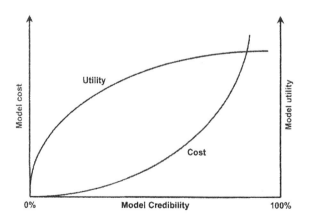

Figure 6-5   Relationship between model cost, credibility, and utility[16]

---

[16] Figure 6-5 is adapted from [6], which itself cites [29] and [28]; the figure also appears in [9].

In Figure 6-5, model credibility increases along the horizontal axis, where it notionally varies from 0% (no credibility whatsoever) to 100% (fully credible). The two curves show how model credibility as the independent variable relates to model utility (how valuable the model is to its user) and model cost (how much it costs to develop, verify, and validate the model) as dependent variables. The model utility curve shows that model utility increases with model credibility, i.e., a more credible model is thus a more useful one, but that as credibility increases additional fixed increments of credibility produce diminishing increments of utility. In other words, there is a point at which the model sufficient credibility for the application, and beyond that point adding additional credibility through the expenditure of additional resources on development, verification, and validation is not justified in terms of utility gained.

The model cost curve shows that additional credibility results in increased cost, and moreover, additional fixed increments of credibility come at progressively increasing cost. In other words, there is a point at which the model has reached the level of credibility inherent in its design, and beyond that point adding additional credibility can become prohibitively expensive. Simulation project managers might prefer to treat cost as the independent variable and credibility as the dependent variable; Figure 6-5 shows that relationship as well through a reflection of the credibility and cost axes and the cost curve. The reflected curve, showing credibility as a function of cost, increases quickly at first and then flattens out, suggesting that the return in credibility reaches a point of diminishing returns for increments of cost.

It is up to the simulation project manager to balance the project's requirements for credibility (higher for some applications than for others) against the resources available to achieve it, and to judge that utility that will result from a given level of credibility.

Verification and Validation Challenges

As with any complex process, the practice of verification and validation involves challenges. In this section five important verification and validation challenges will be presented.

Managing Verification and Validation

Verification and validation activities must be done at the proper times within a simulation project, and they must be allocated sufficient resources. As discussed earlier, the project artifacts to be compared in verification and validation become available at different times in the project; e.g., the executable model will be available for comparison to the model before the results are available for comparison to the simuland. Because of this, the "as soon as possible" guideline implies that verification and validation should be done over the course of the project.[17]

---

[17] Indeed, in the list of 15 principles of verification and validation in [6], this is the first.

Unfortunately, in real-world simulation projects, verification and validation are all too often left to the end of simulation projects. This has at least two negative consequences. First, problems are found later than they might have been, which almost always makes them more difficult and cost more to correct. Second, by the end of the project schedule and budget pressures can result in verification and validation being given insufficient time and attention [7]. All too often, verification and validation are curtailed in simulation projects facing impending deadlines, producing a model that has not be adequately validated and results that are not reliable. The project manager must ensure that verification and validation are not skipped or shortchanged.

Interpreting Validation Outcomes

For all but the simplest models, it is rarely correct to simply claim that a model is "valid" or has "been validated".[18] Most models with practical utility are likely to be sufficiently complex that a more nuanced and qualified description of its validity is likely to be necessary. For such models, validation is likely to show that the model is valid (i.e., is accurate enough to be usable) under some conditions, but not under others. A model might be valid for inputs within a range of values, but not outside those values; for example, the simple gravity model presented earlier could be considered valid for small initial height values, but not height values so large that the distance implied by the height value would cause the acceleration due to gravity to be noticeably less. Or, a model might be valid while operating within certain parameters, but not outside those; for example, an aircraft flight dynamics model could be valid for subsonic velocities, but not for transonic or supersonic. Finally, the accuracy of a model could be sufficient for one application, but not for another; for example, a flight simulator might be accurate enough for entertainment or pilot procedures training, but not accurate enough for official aircraft type qualification. In all of these cases, describing the model as "valid" without the associated conditions is misleading and potentially dangerous. The person performing the validation must take care to establish and document the bounds of validity for a model, and the person using the model must likewise take care to use it appropriately within those bounds of validity.

Combining Models

Models are often combined to form more comprehensive and capable models.[19] For example, a mathematical model of an aircraft's flight dynamics might be combined with a visual model of its instrument panel in a flight simulator. The means of combination are varied, including direct software integration, software architectures, interface standards, and networked distributed simulation; these details of these methods

---

[18] The second principle of verification and validation in [6] makes this point.

[19] In the literature, the terms *composed* or *integrated* are often used instead of *combined*.

are beyond the scope of this chapter (see [13] or [14] for details). However, the combination of models introduces a validation challenge. When models that have been separately validated are combined, what can be said about the validity of the combination? Quite frequently, in both the research literature and in the practice of modeling and simulation (especially the latter), it is assumed that the combination of validated models must be valid, and that validation of the combination is unnecessary or redundant. In fact, a combination of validated models is not necessarily valid; this has been recognized for some time and more recently formally proven [6, 15]. Consequently, when models are combined, a complete validation approach will validate each submodel separately (analogous to unit testing in software engineering) and then validate the combined composite model as a whole (analogous to system testing).[20]

Data Availability

Validation compares the model's results to the simuland's behavior. To make the comparison, data documenting observations of the simuland is required. For some simulands such data may be available from test results, operational records, or historical sources or the data may be easily obtainable via experimentation or observation of the simuland. However, not all simulands are well documented or conveniently observable in an experimental setting. A simuland may be large and complex, such as theater-level combat or national economies, for which history provides only a small number of observable instances and data sufficiently detailed for validation may not have been recorded. Or, a simuland may not actually exist, such as a proposed new aircraft design, for which observations of the specific simuland will not be available. The availability (or lack thereof) of reliable data documenting observations of the simuland will often determine which validation method is used. This is one reason for the relatively frequent application of face validation; subject matter experts are often available when data necessary to support a quantitative validation method, such as statistical analysis, are not.

Identification of Assumptions

Assumptions are made in virtually every model. The example gravity model assumes proximity to the earth's surface. A flight simulator might assume calm, windless air conditions. The assumptions themselves are not the challenge; rather, it is the proper consideration of them in verification and validation that is. Models that are valid when their underlying assumptions are met may not be outside when they are not, which implies that the assumptions, which are characteristics of the model, become conditions on the validity of the model. This is not a small issue; communicating all of the assumptions of a model to the person perform-

---

[20] In [6], there are five levels of testing. The separate validation of submodels is level 2, *submodel (module) testing*, and the combined validation of the composite model is level 4, *model (product) testing*.

ing the validation is important, but the developer of a model can make many assumptions unconsciously and unrealized, and identifying the assumptions can be difficult.[21]

VERIFICATION AND VALIDATION EXAMPLES

The practice of verification and validation is a varied as the subjects of the models involved. In this section several actual examples of verification and validation are presented.

Verification Using Comparison Testing

Comparison testing is a dynamic verification method than can be used when multiple models of the same simuland are available. The models are executed with the same input and the models' results are compared. Differences between the two sets of results suggest possible accuracy problems [6].

Comparison testing was used to verify C^2PAT, a queuing theory-based closed form model of command and control systems [16]. C^2PAT is intended to allow analysis and optimization of command and control system configurations. C^2PAT models command and control systems as a network of nodes representing command and control nodes connected by edges that represent communications links. A set of cooperating agents, known as servers, located at the network nodes exchange information via the connecting links. When a unit of command and control information, known as a job, arrives at a node, that node's server processes it and passes information to connected nodes. Processing time at a node is determined by exponential and non-exponential distributions, and various priority queuing disciplines are used to sequence jobs waiting at nodes to be served. Preemption in the job queues, something likely to happen in command and control systems, is also modeled. C^2PAT models the dynamic response of the command and control system in response to time-varying job arrival rates, determining response and delay times for both individual nodes and for threads of job execution. C^2PAT's queuing theory-based model is analytic, computing values for the parameters of interest in the system using closed-form queuing theory equations.

To verify C^2PAT, two additional distinct versions of the model were implemented, a time-stepped model written in a discrete event programming environment and an event-driven model written in a general purpose programming language. Unlike C^2PAT, the comparison models were stochastic and numerical, simulating the flow of information in the command and control system over time using random draws against probability distributions describing service and delay times. A series

---

[21] This point is made more formally in [30], where the assumptions upon which a model's validity depends are captured in the idea of *validation constraints*.

of progressively more complicated test networks were developed and the three models were run on them. Differences between them were identified and analyzed, and revisions were made to the appropriate model.

The direct comparison of results was quite effective at discovering and focusing attention on potential model accuracy problems. The verification process proved to be iterative, as each comparison would reveal issues to resolve, often leading to revisions to one or another of the models, necessitating an additional run and comparison. The verification effort was ultimately essential to successful modeling in C^2PAT [16].

Validation Using Face Validation

As defined earlier, face validation is an informal validation method wherein expert observers subjectively compare simulation results with their knowledge of the behavior of the actual simuland. Differences between the simulation results and the experts' expectations may indicate model accuracy issues.

Face validation was used to validate the Joint Operations Feasibility Tool (JOFT), a model of military deployment and sustainment feasibility developed by the U.S. Joint Forces Command Joint Logistics Transformation Center [17]. JOFT is intended to be used to quickly assess the feasibility of deployment transportation for military forces to an area of operations and sustainment for those forces once they have been transported. Using JOFT has three basic stages. First, based on a user-input list of military capabilities required for the mission, JOFT identifies units with those capabilities, and the user selects specific units. Second, the user provides deployment transportation details, such as points of embarkation and debarkation, type of transportation lift, and time available. JOFT then determines if the selected force can be deployed within the constraints, and provides specific information regarding the transportation schedule. Third, given details of the initial supplies accompanying the units, the supplies available in the area of operations, a rating of the expected operations tempo and difficulty of the mission, and a rating of the expected rate of resupply, JOFT calculates the sustainment feasibility of the force and identifies supply classes for which sustainment could be problematic.

JOFT was assessed using a highly structured face validation by a group of logistics subject matter experts. Several validation sessions, with different groups of experts participating in each session were conducted, all with this procedure:

1. The procedure and intent for the assessment session was explained.

2. The experts were given a tutorial briefing and a live demonstration of the JOFT software.

3. The experts used JOFT hands-on for two previously developed planning scenarios.

4. The experts provided written feedback on the JOFT concepts and software.

A total of 20 experts participated in the assessment in four different sessions. Collectively, they brought a significant breadth and depth of military logistics expertise to the validation. Of the 20 experts, 17 were currently or had previously been involved in military logistics as planners, educators, or trainers. The remaining three were current or former military operators, i.e., users of military logistics.

The experts' assessments were secured using questionnaires. Categories of questions asked the experts to validate JOFT's accuracy and utility in several ways:

1. Suitability for its intended uses (e.g., plan feasibility "quick look" analysis).

2. Accuracy of specific features of the JOFT model (e.g., resource consumption rate).

3. Utility within the logistical and operational planning processes.

Overall, the experts saw JOFT's greatest strengths as its simplicity, ease and speed of use, and integration of force selection with deployment and sustainment feasibility. They saw JOFT's greatest weaknesses as its lack of accurate data, insufficient detail in calculations, and overlap of capabilities with existing tools.

The validation process was quite effective at identifying both strengths and weaknesses in JOFT; this was due in large part to the high degree of structure and preparation used. The test scenarios were carefully designed to exercise the full range of JOFT functionality and the questionnaires contained questions that served to secure expert assessment of JOFT's validity in considerable detail. The effectiveness of face validation as a validation method is often significantly enhanced by such structure.

Validation Using the Turing Test

The **Turing Test** is an informal validation method well suited to validating models of human behavior first proposed as a means to evaluate the intelligence of a computer system [18]. As conventionally formulated, a computer system is said to be intelligent if an observer cannot reliably distinguish between system-generated and human-generated behavior at a rate better than chance. When applied to the validation of human behavior models, the model is said to pass the Turing Test and to be valid if expert observers cannot reliably distinguish between model-generated and human-generated behavior. Because the characteristic of the system-generated behavior being assessed is its degree of indistinguishability from human-generated behavior, this test is clearly directly relevant to the assessment of the realism of algorithmically generated behavior, perhaps even more so than to intelligence as Turing originally proposed.

The Turing Test was used to experimentally validate the Semi-Automated Forces (SAF) component of the SIMNET distributed simulation, a networked simulation used for training tank crews in team tactics by immersing them in a virtual battlefield [19]. The SAF component used algorithms encoding tactical doctrine to automatically generate and control autonomous battlefield entities, such as tanks and helicopters. In the validation, two platoons of soldiers fought a series of tank battles in the SIMNET virtual battlefield. In each battle, one of the platoons defended a position against attacking tanks controlled by the other platoon of soldiers, the automated SAF system, or a combination of the two. Each platoon of soldiers defended in two different battles against each of the three possible attacking forces, for a total of twelve battles. The two platoons of soldiers had no contact with each other before or during the experiment outside of their encounters outside of the virtual battlefield. Before the experiment, the soldiers were told that the object of the test was not to evaluate their combat skills but rather to determine how accurately they could distinguish between the human and SAF attackers. When asked to identify their attackers after each battle, they were not able to do so at a rate significantly better than random chance. Thus the SIMNET SAF system was deemed to have passed the Turing Test to be validated [19].

Although the Turing Test is widely advocated for use, and used, for validating models of human behavior, its efficacy for that application is critically examined in [20], where it is argued that the Turing Test cannot be relied upon as the sole means of validating a human behavior generation algorithm. Examples are given that demonstrate that the Turing Test alone is neither necessary nor sufficient to ensure the validity of the algorithm. However, if attention is given to the questions of who the appropriate observers are and what information about the generated behavior is available to them, a well-designed Turing Test can significantly increase confidence in the validity, especially in terms of realism, of a behavior generation algorithm that passes the test. An example of such an application of the Turing Test, with results analyzed by an appropriate statistical hypothesis test, was used as a complement to another validation method in evaluating a computer model of decision making by military commanders [21].

Validation Using Hypothesis Testing

Hypothesis testing is a broadly useful statistical technique that can be used as a dynamic validation method. In general, hypothesis testing seeks to determine if a conjecture about some aspect of the phenomenon or population being studied is strongly supported by sample data [22]. There are a number of different statistical hypothesis tests, applicable in different circumstances.

Statistical hypothesis testing was used to validate a model of human decision making [21]. The decision model to be validated was based on recognition-primed decision making, a psychological model of decision making by experienced human decision makers, and implemented using multi-agent software techniques [23]. The goal of the model, known as RPDAgent, was not to make theoretically optimum decisions, but rather to mimic the decisions of human decision makers. In particular, the model was challenged to model the decisions of senior military commanders at the operational level of warfare.

A test scenario based on an amphibious landing was devised to validate the .model. The scenario included four specific decisions, or decision points; one of four landing locations, one of four landing times, one of three responses to changes in enemy deployment, and one of two responses to heavy friendly casualties had to be selected. The choices were numerically coded to allow the computation of mean decisions. Relevant military information about the decision choices was included with the scenario and used by the decision makers. Thirty human military officers were asked to make selections for each of the four decision points within the test scenario. The RPDAgent model has stochastic elements, so 200 replications of 30 sets of the four decisions were generated using the model. For each of the four decision points, the distribution of the human decisions was compared with the distribution of RPDAgent's decisions choice by choice; the number of the humans selected a particular choice was compared to the mean number of times the choice was selected by the model over the 200 replications.

The statistical method employed for the comparison, equivalency testing, can be used to determine if the difference between two distributions is insignificant, as defined by a maximum difference parameter [24]. For the RPDAgent validation, the maximum difference was set to 20%, a value judged to be small enough for realism but large enough to allow for reasonable human variability.

The hypotheses and statistics used for the equivalency testing were

Test 1 hypotheses

$$h_0 : \bar{X} - \mu \leq \delta_1$$
$$h_a : \bar{X} - \mu > \delta_1$$

Test 2 hypotheses

$$h_0 : \bar{X} - \mu \geq \delta_2$$
$$h_a : \bar{X} - \mu < \delta_2$$

Test 1 test statistic

$$t_1 = \frac{(\bar{X} - \mu) \leq \delta_1}{S_{\bar{X} - \mu}}$$

Test 2 test statistic

$$t_2 = \frac{(\bar{X} - \mu) \leq \delta_2}{S_{\bar{X} - \mu}}$$

where

$\bar{X}$ = mean model decision response (times a choice was selected)

μ = human decision response (times a choice was selected)

δ_1 = lower limit of the equivalency band

δ_2 = upper limit of the equivalency band

Two statistical hypothesis tests are needed in equivalency testing. For this application, test 1 determines if the model mean is less than or equal to the human mean allowing for the equivalency band; rejecting the null hypothesis shows that it is not. Test 2 determines if the model mean is greater than or equal to the human mean, allowing for the equivalency band; rejecting the null hypothesis shows that it is not. Rejecting both null hypotheses shows that the distributions are equivalent within $\pm\delta$. For both tests a one-tailed Student t-test was used with $\alpha = 0.05$, giving a critical value for the test statistic $t = 1.645$.

The model and human means were compared in this manner for each of the choices for each of the four decision points in the test scenario. In every comparison, the calculated test statistics exceeded the critical value for both test 1 and test 2, leading to the rejection of both null hypotheses, thereby supporting the conclusion that the model and human decision means were equivalent within the difference parameter.

Use of a statistical validation technique is not always possible. To do so, the data needed to support it must be available and the appropriate technique must be selected and properly employed. However, when these conditions are met, such methods can be quite powerful, as this example shows.

Validation Using Regression Analysis

Regression analysis is another multi-purpose statistical technique that can be used as a dynamic validation method. In general, regression analysis seeks to determine the degree of relatedness between variables, or to determine the extent that variation in one variable is caused by variation in another variable [22].

Regression analysis was used to validate a model of spacecraft mass [25]. The Spacecraft Propulsion System Sizing Tool (SPSST) model predicts the mass of the propulsion system of automated exploration spacecraft. The propulsion system can account for as much as 50% of a spacecraft's total mass before liftoff. The model is intended to support engineering trade studies and provide quick insight into the overall effect of propulsion system technology choices on spacecraft mass. The mass prediction is calculated using physics-based equations and engineering mass estimating relationships. Inputs to the model include mission profile parameters, such as velocity change and thermal environment, and selected options for nine subsystems, including main propellant tanks, main propellant pressure system, and main engines. The model outputs predicted mass for the overall spacecraft propulsion system, both with and without propellant (called *wet mass* and *dry mass*, respectively) as well as for the spacecraft subsystems, for the given mission.

To validate the SPSST model, mass, mission, and subsystem option data were collected for twelve existing spacecraft, including Mars Odyssey, Galileo, and Cassini. The SPSST model was used to predict the propulsion system and subsystem masses for these spacecraft, given the characteristics of these spacecraft and their missions as input. The resulting model predicted values for wet mass, dry mass, and subsystem mass were compared with the actual spacecraft mass values using linear regression. When the twelve pairs of related predicted and actual wet mass values were plotted as points, they fell quite close to line, suggesting accuracy in the model. The computed coefficient of regression statistic was $R^2 = 0.998$; the statistic's value close to 1 confirmed the model's accuracy for this value. Similar results were found for dry mass, though here the statistic's values was somewhat lower, $R^2 = 0.923$. On the other hand, the subsystem predictions were not nearly as consistent; the model did well on some subsystems, such as propellant tanks, and not as well on others, such as components.

Overall, linear regression provided a straightforward and powerful validation method. When model results provide values that can be directly paired, linear regression may be considered.

CONCLUSION

The outcome of verification and validation is normally not a determination that the model is entirely correct or entirely incorrect [6]. Rather, verification and validation serve to determine the degree of accuracy a model has, and the ranges or types of inputs within which that accuracy is present.

Verification and validation are, as already stated, essential prerequisites to the credible and reliable use of a model and its results. Shortchanging verification and validation is false economy, as the consequences of using an invalid model can, in some circumstances be dire.

KEY TERMS

Simuland	accreditation	predictive validation
model	face validation	formal verification and validation
conceptual model	data analysis	inductive assertions
executable model	cause-effect graphing	predicate calculus
simulation	dynamic verification and validation	Turing Test
verification	sensitivity analysis	regression analysis
validation	inspection	

REFERENCES

1. FORD T. Helicopter simulation. *Aircraft Engineering and Aerospace Technology* 1997;69(5):423–427.
2. KESSERWAN N. Flight Simulation [thesis]. Montreal (Canada): McGill University; 1999.
3. RUMBAUGH J, JACOBSON I, BOOCH G. *The Unified Modeling Language*. Reading, MA: Addison-Wesley; 1999.
4. BANKS J, CARSON JS, NELSON BL. *Discrete-Event System Simulation*, 2nd Ed. Upper Saddle River, NJ: Prentice Hall; 1996.

5. U.S. Department of Defense, "Modeling and Simulation," DOD Instruction 5000.61, 1996.
6. BALCI O. Verification, validation, and testing. In: Banks J, ed. *Handbook of Simulation: Principles, Methodology, Advances, Applications, and Practice*. New York: John Wiley & Sons; 1998. pp. 335–393.
7. BALCI O. Verification, validation, and accreditation. Proceedings of the 1998 Winter Simulation Conference; 1996 December 13–16; Washington DC; pp. 41–48.
8. JACOBY SLS, KOWALIK JS. *Mathematical Modeling with Computers*. Englewood Cliffs, NJ: Prentice-Hall; 1980.
9. SARGENT RG. Verification, validation, and accreditation of simulation models. Proceedings of the 2000 Winter Simulation Conference; 2000 December 10–13; Orlando, FL; pp. 50–59.
10. COHEN ML, ROLPH JE, STEFFEY DL. *Statistics, Testing, and Defense Acquisition, New Approaches and Methodological Improvements*. Washington, DC: National Research Council, National Academy Press; 1998.
11. MILLER DR. Sensitivity analysis and validation of simulation models. *Journal of Theoretical Biology* 1974;48(2):345–360.
12. BALCI O, SARGENT RG. A methodology for cost-risk analysis in the statistical validation of simulation models. *Communication of the ACM* 1981;24(4):190–197.
13. DAVIS PK, ANDERSON RH (RAND National Defense Research Institute, Santa Monica CA). Improving the Composability of Department of Defense Models and Simulations; 2003.
14. WEISEL EW, PETTY MD, MIELKE RR. A Survey of Engineering Approaches to Composability. Proceedings Of the Spring 2004 Simulation Interoperability Workshop; 2004 April 18–23; Arlington VA. pp. 722–731.
15. WEISEL EW, MIELKE RR, PETTY MD. Validity of Models and Classes Of Models in Semantic Composability. Proceedings Of the Fall 2003 Simulation Interoperability Workshop; 2003 September 14–19; Orlando, FL. pp. 526–536.
16. TOURNES C, COLLEY WN, UMANSKY M. C²PAT, a Closed-Form Command and Control Modeling and Simulation System. Proceedings of the 2007 Huntsville Simulation Conference; 2007 October 30–November 1; Huntsville, AL.
17. BELFORE LA, GARCIA JJ, LADA EK, PETTY MD, QUINONES WP. Capabilities and Intended Uses of the Joint Operations Feasibility Tool. Proceedings of the Spring 2004 Simulation Interoperability Workshop; 2004 April 18–23; Arlington, VA. pp. 596–604.
18. TURING AM. Computing machinery and the mind. *Mind* 1950;59(236):433–460.
19. WISE BP, MILLER D, CERANOWICZ AZ. A Framework for Evaluating Computer Generated Forces. Proceedings of the Second Behavioral Representation and Computer Generated Forces Symposium; 1991 May 6–7; Orlando, FL. pp. H1–H7.
20. PETTY MD. The Turing Test as an Evaluation Criterion for Computer Generated Forces. Proceedings of the Fourth Conference on Computer Generated Forces and Behavioral Representation; 1994 May 4–6; Orlando, FL. pp. 107–116.
21. SOKOLOWSKI JA. Enhanced Decision Modeling Using Multiagent System Simulation. *Simulation* 2003;79(4):232–242.
22. BHATTACHARYYA GK, JOHNSON RA. *Statistical Concepts and Methods*. New York: John Wiley & Sons; 1977.
23. KLEIN G. Strategies of decision making. *Military Review* 1989;69(5):56–64.
24. ROGERS JL, HOWARD KI, VESSEY JT. Using significance tests to evaluate equivalence between two experimental groups. *Psychological Bulletin* 1993;113(3):553–565.
25. BENFIELD MPJ. Advanced Chemical Propulsion System (ACPS) Validation Study. Unpublished presentation. University of Alabama in Huntsville, November 28, 2007.
26. U.S. Department of Defense. VV&A Recommended Practices Guide. RPG Build 3.0 September 2006. Available at http://vva.dmso.mil/. Accessed 2008 March 13.
27. ROYCE W. *Software Project Management, A Unified Framework*. Reading, MA: Addison-Wesley; 1998.

28. SARGENT RG. Verifying and Validating Simulation Models. Proceedings of the 1996 Winter Simulation Conference; 1996 December 8–11; Coronado, CA. pp. 55–64.

29. SHANNON RE. *Systems Simulation: The Art and Science*. Upper Saddle River, NJ: Prentice Hall; 1975.

30. SPIEGEL M, REYNOLDS PF, BROGAN DC. A Case Study of Model Context for Simulation Composability and Reusability. Proceedings of the 2005 Winter Simulation Conference; 2005 December 4–7; Orlando, FL. pp. 436–444.

Part Three

Practical Domains

Chapter 7

Uses of Simulation

Tuncer I. Ören

Simulation is like a gem: it is multifaceted[1]
—Tuncer Ören, 1984

INTRODUCTION

Two complementary approaches can be useful in discussing the uses of simulation: 1) to provide a list of application areas of simulation—with or without additional clarifications—and 2) to provide a systematic basis for possible uses of simulation. For the first approach, a list of application areas of simulation can be compiled, for example, from simulation conference announcements and/or proceedings. Even with 10–15 conferences, you can prepare an informative and useful list. In a systematic basis, in turn, you can present categories of uses in a comprehensive way. In this approach, the user can decide the applicability of simulation even in novel application areas. Both of the approaches have their merits. Since this textbook is for a multidisciplinary audience at the undergraduate or beginning graduate level, a list of application areas may give you a good idea about the pervasiveness of simulation in large number of application areas. For this reason, some lists of application areas of simulation are given; in addition to the important application areas of modeling and simulation (M&S) introduced and explained in Chapter 1. After well over 40 years of involvement, I still enjoy discovering new frontiers in M&S. I would hope that, similarly, this chapter might open new vistas for simulationists of many years of involvement as well as for students new in exploring the domain. A systematic

---

[1] In 1984 Dr. Tuncer Ören wrote the foreword to the book *Multifaceted Modelling and Discrete Event Simulation*, by Bernard P. Zeigler (Academic Press London, England). In it he noted: *Simulation is like a gem: it is multifaceted.*

Principles of Modeling and Simulation: A Multidisciplinary Approach, Edited by John A. Sokolowski and Catherine M. Banks.

basis for the uses of simulation requires a presentation of the possibilities that simulation offers [1–3]. A top-down systematic discussion may start with elaboration on the many facets of simulation.

THE MANY FACETS OF SIMULATION

As it is stressed in the opening quotation, simulation has many facets: some are essential for the discipline of M&S; and some have been used in the English language for a long time but are not part of the technical aspects of simulation. The word **simulation** has been used in English since the year 1340 [4]. It is related with Latin word *simulacrum* (plural: simulacra and also simulacrums), which is from the Latin *simulare*, "to make like," from similes, "like" [5]. As a nontechnical term, simulation means something similar to but not the real thing; hence it means a *representation*. Depending on the goal of the representation, simulation may also mean imitation, pretension, false appearance, feigning (i.e., pretending with intention to deceive), and fake. An example is the term *simulated pearl* to designate *cultivated pearl*, which is not real pearl but an imitation, similar to a real one. A few years ago when I was in southern France I asked a travel agent for the details of a trip to a Mediterranean island, and the travel agent replied in French: "Let's have a simulation." What she meant was "let's *pretend* that I am going to issue a ticket for your trip and get the information, without issuing the ticket at the last step." Hence, with her *pretending* to issue a ticket, I got the information but not the ticket for the trip. However, in technical parlance we do not use the term simulation in this sense.

As a technical term, simulation has two major aspects: *experiments* and *experience*. Accordingly, simulation offers a very rich paradigm to *perform experiments* with dynamic models; and, using representations of real systems, to *provide experience* either for entertainment or for training to develop and/or enhance three types of skill, i.e., *motor skills, decision-making skills,* or *operational skills.* The last sentence is indeed loaded and needs to be elaborated on. For this purpose, let's enumerate several aspects of simulation and give separate and relevant definitions of it. Use of a representation of reality—whether existing or yet to be engineered, or purely hypothetical—instead of the real system itself is the prominent characteristic of all technical meanings of simulation.

In this chapter, we concentrate on the technical meanings of simulation. It is worth knowing that American sociologist Daniel Bell stressed the pivotal importance of technical aspects of simulation in post-industrial (or information, or knowledge) societies. This view was presented to simulationists in the 1970s: As Bell, who explains that in post-industrial societies the methodologies are abstract theories: models, simulation, decision theory, and system analysis, states it: "The key political problems in a post-industrial society are essentially elements of science policy" [6, 7]. It is a very satisfactory state of affairs that, at the beginning of the 21st century, simulation is accepted by the U.S. House as a Nationally Critical Technology [8]. It would be wise for many other countries to follow suit to benefit from the advantages of simulation. In the sequel, both technical and nontechnical meanings are provided; however, the latter category is for the sake of completeness.

Technical meanings

Experimentation aspect of simulation:

(Simulation is performed for decision support, understanding, and education.)

1. Simulation is goal-directed experimentation with dynamic models. A dynamic model has time-varying characteristics and can be mathematical, symbolic, or physical.

Experience aspect of simulation:

(Simulation is performed either to provide experience for entertainment or to enhance skills.) Hence, two additional definitions of simulation follow:

2. Simulation is use of a representation of a real system to *provide experience* for *entertainment.*

3. Simulation is use of a representation of a real system to *provide experience* for *training* to develop and/or enhance three types of skill, i.e., motor skills, decision-making skills, or operational skills." For the last three categories of skills, associated three types of simulation are virtual, constructive, and live simulations, respectively.

4. Within the perspective of using a representation (a model) of an existing, desired, or a hypothetical system, to provide experience, simulation also means something similar, an imitation, but not the real thing, e.g., simulated battleground and simulated forces.

5. Hence, a combined and concise definition of simulation is:
Simulation is goal-directed experimentation with dynamic models or use of a representation of a real system to *provide experience* for *entertainment* or for *training* to develop and/or enhance three types of skill, i.e., motor skills, decision-making skills, or operational skills.

Nontechnical meanings

6. Simulation is something similar, an imitation, but not the real thing, e.g., simulated pearl and simulated leather. To contrast this meaning with the meaning given in (4), we have to note that in this meaning, the products (e.g., pearl or leather) are not used for any type of experience.

7. Simulation denotes pretence and not the real process. Consider the travel agent who is asked to provide the price of a trip.

8. Simulation denotes feigning (i.e., pretending with intention to deceive), illusion, phantasm, and fake. However, a more specific term in English is dissimulation and to dissimulate.

For the sake of completeness, we should also mention that Baudrillard, a post-modernist French philosopher, argued that in a post-modern society, most of the perceptions are not representations of the real entities and that simulation is false representation and as such successively,

a. It is the reflection of a basic reality.

b. It masks and perverts a basic reality.

c. It masks the absence of a basic reality.

d. It bears no relation to any reality whatever: it is its own pure simulacrum [10].

One may also note the following: "While his work on simulation and the postmodern break from the mid-1970s into the 1980s provides a paradigmatic postmodern theory and analysis of postmodernity that has been highly influential, and that despite its exaggerations continues to be of use in interpreting present social trends, his later work is arguably of more literary interest. Baudrillard thus ultimately goes beyond social theory altogether into a new sphere and mode of writing that provides occasional insights into contemporary social phenomena and provocative critiques of contemporary and classical philosophy and social theory, but does not really provide an adequate theory of the present age" [11].

9. Finally, a clarification of the term *emulation*. Emulation (to emulate, emulated, emulating, emulator) has technical and nontechnical meanings. In its technical meaning, "emulation is use of a system, device, or software in lieu of another one." In its nontechnical meaning, "emulation is behaving like another person."

Further clarifications on different aspects of simulation as well as its perceptions as several types of knowledge processing can be found in Ören [2, 3]. This section is adapted from them.

EXPERIMENTATION ASPECT OF SIMULATION

Experiment or **experimentation**, as a noun, is 1) a test or a procedure done in a controlled environment for the purpose of gathering observations, or facts; demonstrating known truth; examining the validity of a hypothesis, or a theory; or 2) the process of conducting such a procedure. As a verb, it means to carry out such a test or procedure.

Experimentation is a key scientific concept and "simulation has an important place in *philosophy of science*: 'Experimentation' is one of the key concepts in scientific thinking since Francis Bacon (1561–1626) who advocated it in 1620 in his Novum Organum (New Instrument). Bacon's work was a categorical departure from and reaction to 'Organon' (the Instrument) which was the title of logical works of Aristotle (384–322 B.C.) which itself had an 'unparalleled influence on the history of Western thought'" [12]. "One of the superiorities of simulation over real-system experimentation is that simulation is possible in all cases when experimentation on real system cannot be performed. Furthermore, in simulation experimental conditions can include cases that cannot and should not be performed on real systems" [1].

In experimentation, you can observe either the sequence of values of some variables of interest or the sequence of structural changes of a model. The sequence of values of a variable is also named a trajectory of that variable. Hence, when the aim is to observe trajectory of a variable or trajectories of some variables of interest to the observer, the simulation is called a **trajectory simulation**. Most simulation studies for experimentation purposes are trajectory simulations. However, another possibility is to observe the evolution of the structure of a system model; this is the observation of the sequence of its structures over time. For example, in crystalliza-tion of a fluid metal, you may observe the formation and growth of crystals. Another example is the growth of plants. The arrangement of leaves on a stem follow well-determined and quite interesting formula studied as phyllotaxy [13]. Phyllotaxy and L-systems (Lindenmayer systems) are useful for the simulation of plant growth, which is another example for structural simulation. An example of simulation model-ing of plants and plant ecosystems is by Prusinkiewicz [14]. In the the following discussion we will elaborate on types of systemic observations of a model and types of decision making relevant to simulation.

Systemic Possibilities to Observe a Model

It is a classical knowledge that models can be observed in three different ways [15]. The changes of a system can be described through changes of its states. Hence, as seen in Table 7-1, for three types of system problems, output, state, or input variables can be observed given the two other types of variables.

Analysis Problems

In **analysis problems**—as it is the case in scientific problems—a system of interest exists and often needs to be understood. Hence, simulation can be used to test several models until one finds one that would replicate—with an acceptable tolerance—the trajectory and/or structural trajectory of the system under investigation under similar experimental conditions. Therefore, a model of the system would necessitate

Table 7-1   The Three Types of System Problems in an Experimentation

Type of system problem	Given			Observe
	inputs	states	outputs	
Analysis	inputs	states		outputs
Design	inputs		outputs	states
Control		states	outputs	inputs

appropriate ways to represent its state and the evolution of it through time. Usually, this may be through state variables. In a *trajectory simulation*, given the state transition functions and initial values of state variables, one can stimulate the model with a sequence of values of input variables (or input trajectories) to observe a sequence of values of state variables. With appropriate output functions the values of output variables can also be computed based on the thus computed state variables. In **structural simulation** sequences of structures of the system of interest can be computed based on the initial structure and the input trajectories.

Almost always the experiment starts at time t0 t0_legend and continues until t_f; where $t_f > t_0$. t_f can be specified before the beginning of the experiment or recorded when a *state condition* (a condition that depends on either state variable(s) or state structure of the model) is satisfied. It is also possible—though not used often— to start with a final condition and let time flow backwards. This is called **retro- simulation** (or backward simulation or reverse simulation), which can be used to make retrodictions, i.e., "predictions" about the past [16].

Design Problems

In *design problems*—as it is the case of engineering problems—given a system design specification, one would like to design and then implement a system that would satisfy its technical specifications. Building a system would be equated, in system theoretic terminology, to specifying its states as well as its state transitions by specifying and implementing its state transition function. Hence, from a simulation- ist perspective, a design problem can be conceived of developing a model and testing it under well-defined experimental conditions to see whether its trajectory and/or structural behavior would match the desired behavior as specified in the technical specification under same experimental conditions.

Control Problems

In *control problems*, a system may exist; hence, its state is known and can be repre- sented by a model. The problem is to find appropriate sequence of values of input variable(s) for the system to exhibit desired output. Hence, from a simulationist point of view, the control problem can be reduced to, using a model of the system, finding sequence of input variables that would cause the model to generate desirable output trajectories. Hence, the input-output behavior of the system can be simulated until one finds appropriate input segments that produce desired output. In another section, we will see in more detail how simulation can be used in control problems.

Types of Uses of Simulation for Decision Support

Types of uses of simulation for decision support are outlined in Table 7-2. More information about the experimentation aspect of simulation can be found in Ören [2, 3].

Table 7-2   Types of Uses of Simulation for Decision Support

- *Prediction* of behavior or performance of the system of interest within the constraints inherent in the simulation model (e.g., its granularity)
- *Evaluation of alternative* models, parameters, experimental and/or operating conditions on model behavior or performance
- *Sensitivity analysis*
- *Engineering design*
- *Virtual prototyping*
- *Planning*
- *Acquisition* (or simulation-based acquisition)
- *Proof of concept*

EXPERIENCE ASPECT OF SIMULATION

The term "experience" has the following meanings [17]:

1. The apprehension of an object, thought, or emotion through the senses or mind
2. Active participation in events or activities, leading to the accumulation of knowledge or skill
3. The knowledge or skill so derived
4. An event or a series of events participated in or lived through
5. The totality of such events in the past of an individual or group
6. To participate in personally

As we have seen in the section on its many facets, simulation can be used either to provide experience for entertainment or to enhance skills for training purposes. Hence, experience aspect of simulation can be summarized with the following definition: "Simulation is the use of a representation of a real system to *provide experience* for *entertainment* or for *training* to develop and/or enhance three types of skill, i.e., motor skills, decision-making skills, or operational skills."

Experience Aspect of Simulation for Entertainment

The term **simulation games** distinguishes games in general from those that have simulation components (namely representation of often hypothetical or real systems). A plethora of simulation games exist with associated clubs, magazines, and books and even career opportunities for designers and developers. However, in this chapter, we concentrate on training as far as simulation-based experiences are concerned.

Experience Aspect of Simulation for Gaining/ Enhancing Skills for Training

As we have seen in the definitions section, "simulation is the use of a representation of a real system to *provide experience* for *training* to develop and/or enhance three types of skill, i.e., motor skills, decision-making skills, or operational skills."

Motor Skills

Virtual (simulated) equipment is used to gain/enhance motor skills needed to use equipment(s). Due to the use of *virtual (simulated) equipment*—namely, a model of the system—this type of simulation is used virtual simulation. The terms *simulator* and *virtual simulator* are used to denote nature of models, i.e., *physical model* of the system or *all software model* of the system (e.g., flight simulator, (virtual) flight simulator). Virtual simulators are all software simulators. The operators of equipments, including military equipments, are trained better on simulators or virtual simulators rather then the real system.

Decision-Making Skills

Simulation software is used to provide experience under controlled conditions necessary for training and education to get knowledge, attitude, and experience to make decisions. This type of simulation is called **constructive simulation**, in military terminology. In civilian terms, it is called **gaming simulation** (or serious games, to distinguish them from simulation games used for entertainment). For example, "war games," "business games," "peace games," and "conflict management games" are all simulation games used to enhance decision-making abilities of the people responsible for corresponding types of problems. In any type of simulation game, people (in military applications, forces), equipment, and environment are represented by models (i.e., imitations and not the real system).

Operational Skills

Live simulation is used for training of both of civilian and military personal. In military applications live simulation, i.e., training to gain competency through experience, is performed with real and virtual (simulated) equipment acting in real environment and with imitated (simulated) ammunition. In live simulation, real people and real equipment are augmented with special sensors to act as target designators. In civilian applications, such as medical training, live simulation provides experience needed for medical purposes.

EXAMPLES OF USES OF SIMULATION

Chapter 1 of this book has several examples of uses of simulation. In this chapter, we will see 1) some classical uses, from the mid-1960s to mid-1970s (most, if not all of the application areas are still active) and 2) some application areas from several symposia of a simulation conference in 2008. Currently, there are over 110 simulation associations, groups, and centers. Exploring their activity areas would point out much more specific application areas. Some associations, groups, and centers have specific application areas such as military, business, health sciences, etc. This would make search of certain types of uses much easier. Furthermore, you as the reader may evaluate the possibilities of several types of applications of simulation based on the background knowledge given about experimentation and experience aspects of simulation.

Some Classical Uses of Simulation

Some uses of simulation (from the proceedings of early Summer and Winter Simulation Conferences are listed in Table 7-3 [18]. The bibliography covers 1227 articles published in over 10,000 pages of the first 13 Summer and Winter Simulation Conferences held during 1967 through 1974. Almost all application areas of simulation given in this table are still active and simulation continues to the advancement of knowledge in these disciplines or application areas.

Some Current Uses of Simulation

In Tables 7-4,a–j, the application areas—as expressed in their call for papers—of the symposia of only one conference (the Spring Simulation Conference) with which the author was involved are listed [19]. Even the application areas of the symposia of one single conference represent the large scope of applicability of simulation.

Simulation is successfully applied in hundreds of other and more specific areas. The best way to explore other application areas of simulation is to visit web sites of over 110 associations, groups, and research centers active in simulation. Several tables are given in Appendix A to facilitate exploration of their activities. An example of the use of simulation in control problems follows.

Use of Simulation in Control Problems

Control systems can benefit from the contribution of simulation in many ways. In a recent excellent habilitation thesis, Santoni covers user/system interfaces for industrial control systems [20]. Uses of simulation—most of which are covered by Santoni—in user/system interfaces of control systems are as follows:

- Use of simulation for the training of operators of industrial control systems
  (under normal as well as abnormal operating conditions; with detailed analyses of performance and severity of mistakes)

Table 7-3 Application Areas of Simulation (from an
unpublished bibliography prepared in the mid-1970s)

Accident (in marine transportation)
Accounting
Aeronautics/Astronautics/Aerospace
Agriculture
Air-sea interaction
Airport
Anatomy (skull)
Applied process
Astronomy
Bio-chemistry/Biology
Bio-engineering/Bio-physics
Bio-medicine (clinical)
Breathing
Boiler
Budget
Cardiovascular system
Chemical process
Chemical reactions
Chemistry
Circuit
Civil engineering
Climate
Communication
Company/Corporate models
Compressor
Computer (hardware, multiprocessing, multiprogrammed)
Computer reliability
Computer software
Cost/Cost effectiveness
Decision
Desalination process
Diagnosis/Fault diagnosis
Diffusion
Disaster planning
Distillation
Distribution
Drug
Earth science
Earthquake
Ecology/Environment
Economics
Electronic components

Table 7-3   *Continued*

Energy
Filters (digital)
Financial models
Flow in manufacturing systems
Flow (fluid, gas)
Forecasting
Food and drug industries
Gambling
Garment
Gasoline plant
Genetics
Geophysics and geology
Health services
Human behavior
Human factors engineering
Human locomotion
Hydrology/Hydraulic systems
Industrial systems
Information retrieval systems
Information systems (management/engineering)
Instrumentation
Inventory
Investment
Ion exchange columns
Job shop
Land use
Legislative/Judicial systems
Logistics
Magnetic anomalies
Maintenance/repair
Management
Manufacturing/Production
Marine engineering
Marketing
Material (including raw material)
Material handling/conveyor/crane systems
Measurement
Mechanical engineering
Mechanical linkages
Metrology
Military (war/weapons, missile)
Music
Network

Table 7-3   *Continued*

Nitrification
Nuclear and general physics
Nuclear engineering
Oceanography
Optical character recognition
Petroleum/Power industries
Planning
Plant sciences
Political sciences/Systems
Pollution (air, water)
Population dynamics—plant and animal
Portfolio
Printing
Public policy
Queuing
Radar
Reactor
Reliability/Maintainability
Reservation systems
Resource
Scheduling
Semiconductor
Social systems
Soil
Stock market/Commodities
Tracking
Traffic control (air)
Transportation (air, helicopter, highway/urban, railway, sea)
Universities
Urban systems (demography, fire department, housing,
  police)
Utilities
Warehouse
Waste treatment
Water (ground/surface)

- Use of simulation for specification and synthesis of command systems
  —Verification of the specification by using two types of simulators:
  —Structural simulator (to verify the rules of construction of command graph)
  —Functional simulator (to automate verification of the interpretation of the
    commands)

Table 7-4a Application Areas of SpringSim'08 ADS'08
(Agent-directed Simulation Symposium)

- Simulation modeling of agent technologies at
  —the organization,
  —interaction (e.g., communication, negotiation,
  coordination, collaboration) and
  —agent level (e.g., deliberation, social agents,
  computational autonomy)
- Application of agent simulations in various areas such as
  biology, business, commerce, economy, engineering,
  environment, individual, group, and organization behavior,
  management, simulation gaming/training, social systems
- Conflict management simulation with holonic agents

Table 7-4b Application Areas of SpringSim'08 ANSS'08
(Annual Simulation Symposium)

Artificial intelligence in simulation
Cognitive Modeling and Simulation
Energy-aware Schemes for Wireless Networks
Network Modeling and Simulation
Neural Network Models and Simulation
Parallel and Distributed Simulation
Reliability and Maintenance Models
Routing and Mobility Management in Networks
Simulation-based Performance Analysis
Simulation-based Software Performance
Simulation of Agent Systems
Simulation of Client-Server Systems
Simulation of Clusters and Grids
Simulation of Distributed Systems and Databases
Simulation of Large-scale Systems
Simulation of Multimedia Applications and Systems
Simulation of Multiprocessor Architectures
Simulation of Parallel Processing Systems
Simulation of Real-time Systems
Simulation of Sensor Networks
Simulation of Wireless and Mobile Networking and
  Computing
Simulation of Wireless Systems
Smart Network Design and Traffic Modeling
VLSI Circuit Simulators

Table 7-4c   Application Areas of SpringSim'08 BIS'08
(Business and Industry Symposium)

Automotive Simulation Applications
Decision Analysis
Decision Support Systems
Dynamic Business Process Optimization
Economic, Financial, and Marketing Systems
Environmental Sciences and Applications
Genetic Programming
Global Optimization
Knowledge and Data Systems
Machine Learning
Management Systems
Manufacturing Plants and Systems
Material Sciences and Applications
Mathematical Algorithms/Methodologies
Neural Networks/Fuzzy Logic
Nuclear and Chemical Waste Processing Technology
Operations Research
Power Plant Simulators
Process Modeling
Systems and Process Simulation
Systems Optimization
Transportation Systems
Virtual Reality Systems

Table 7-4d   Application Areas of SpringSim'08 CNS'08
(Communications and Networking Simulation Symposium)

Ad Hoc Networks and Applications
Clusters and Grid Computing & Communications
Communications and Networks
Data Communications and Protocols
Distributed Simulation and Real-Time Systems
Dynamic Data Driven and Sensor Networks
High-Level Architecture for Networking Industry
Large-Scale Networks
Load Balancing and Congestion Control
Network Centric Defense Application
Network Centric Virtual Environments
Network Design and Performance Analysis
Network Security and Management
Traffic Engineering and Measurements
Virtual Environments for Training
Wireless Communications and Mobile Networks

Table 7-4e Application Areas of SpringSim'08 DODAF'08
(DOD Architecture Framework Modeling)

DODAF 2.0
DODAF models (intelligent agents for)
DODAF models (new representations and extensions for)
DODAF-based network simulations
Simulation-based acquisition
Vulnerability assessment (DODAF-based)

Table 7-4f Application Areas of SpringSim'08 HPCS'08
(High Performance Computing and Simulation Symposium)

High-performance/large-scale simulation
Hybrid system modeling and simulation

Table 7-4g Application Areas of SpringSim'08 ICHSS-17
(17th International Conference on Health Sciences
Simulation)

Bioengineering (cardiac devices, regeneration)
Drug development (clinical trials, population analyses)
Epidemiology (disease transmission, treatment costs)
Health policy (nutrition, risk evaluation, smoking cessation)
Health services (resource allocation, care assessment)
Medical decision making (telemedicine, screening,
 treatment)
Medical practice (cancer survival, obesity control)
Microbiological processes (genetic control, AIDS
 pathogenesis)
Pharmacodynamics (dose response, synergy, toxicology)
Physiology (pulmonary function, endocrine systems)

- •Use of simulation to improve the security of industrial systems by studying their human-system interfaces
- •Use of simulation to minimize impact of human error on the security of the controlled systems.

Another important possibility for control systems is the use of simulation to realize **predictive displays** (Figure 7-1). Systems with predictive displays work as follows: along with the controlled system, there is a simulation model of it. The controlled system's state is used to update the state of its simulation model. Furthermore, the simulation model automatically receives values of environmental

Table 7-4h Application Areas of SpringSim'08 MMS'08 (Military Modeling and Simulation Symposium)

- Advanced Concepts and Requirements
  Intelligent systems simulation
- Domains (land, sea, air)
  Avionics, flight control, flight simulation
  Surface and subsurface warfare
  Synthetic environments/virtual realities
  Unmanned vehicles
- Operations and Command and Control
  Airspace management
  Battle field/battle theater simulation
  C4I simulation
  Counterforce operations
  Simulation during operations
- Physical Modeling and Effects
  Ballistics & propellant simulation
  Computational fluid and molecular dynamics
  Impact and penetration modeling
  Lethality, vulnerability survivability
  Structural and solid mechanics modeling
- Research, Development, and Acquisition
  Design, development, and acquisition for new weapons systems and equipment
  Simulation and modeling for acquisition, requirements, and training (SMART)
  Simulation-based acquisition
- Serious Games
  Computer-generated forces
  Game AI for military M&S
  Game-based simulation
  Games in training
- Simulation Support to Operations
  Battlefield visualization
  Course of action analysis
  Electronic performance support systems
  In-theater training
  Mission rehearsal
- Synthetic Environment Development
  Agent-based combat modeling
  Computer-generated forces
- Training, Exercises, and Military Operations
  Assessment
  Embedded training
  Mission planning & rehearsal
  Simulation/exercise integration
  Simulations in training

Table 7-4i Application Areas of SpringSim'08 MSE
(Modeling and Simulation in Education)

Modeling, simulation, and gaming as a teaching tool

Table 7-4j Application Areas of SpringSim'08 SSSS'08
(Symposium on Simulation of System Security)

Intrusion detection and prevention

Mesh networks security

Modeling or simulating vulnerabilities in systems

Models of secure behavior (OS models, policy models,
 access control models, and authentication models)

Network-centric warfare system security

Scientific visualization techniques applied to computer
 security

Secure distributed systems and networks

Sensor networks security

Use of M&S in information security training

Use of M&S in network and system security

Use of M&S in security engineering

Wireless networks security

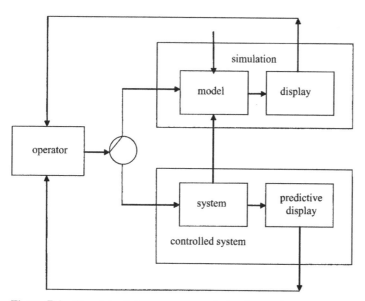

Figure 7-1   Use of simulation to provide predictive displays for controlled systems

variables. Based on the information on the displays of the controlled system, the operator decides on the nature and values of the inputs. However, the operator's inputs go first to the simulator. The values computed by the simulator are displayed in displays that predict the behavior of the model. If the trajectories displayed in the predictive displays are acceptable, then a switch re-routes the operator's inputs to the controlled system.

ETHICS IN THE USE OF SIMULATION

The model used in a simulation study should be representative of the system under investigation for the purpose of the study. This implies two important concepts, namely, validation and verification. **Validation** necessitates appropriateness of the conceptual model to represent the system under investigation. **Verification** is the assurance that the computerization of the conceptual model is done properly without introduction of computational errors. A discussion by Dr. Osman Balci covers several principles of simulation model validation and verification [21].

A word of caution in using simulation: if our actions have serious implications to others, ethical behavior becomes paramount. Both technical uses of simulation, namely providing experience for training purposes and performing experimentation with dynamic models have serious implications [22]. Hence, ethics becomes necessary for sustainable civilized behavior [23]. A Code of Ethics for Simulationists is given by [24]. Further information on ethics for simulationists can be found at the SimEthics website [25].

SOME EXCUSES TO AVOID USING SIMULATION

Two common criticisms of simulation are the claims that it is time-consuming and expensive. For the first criticism, the analogy of the position of a medical doctor in an operating room may be useful. The tools should be previously developed and ready for the operation. Furthermore, the doctor should have the training to use the advanced tools. If the doctor does not have the tools needed at the time of operation, it may be too late for the patient since it would be absurd to ask a tool developer to develop the needed tool(s) just at the time of need. Similarly, simulation model(s) should be ready to be used with or without tailoring at the time of use. Not having the necessary infrastructure for simulation is a deficiency of the potential user rather than a deficiency of the simulation discipline.

Those who would claim that simulation studies are expensive should also consider the cost of not using simulation. Furthermore, in decision support, for example, having an explicit model with its assumptions and usability conditions clearly stated is always better than making impromptu decisions. An explicit model can be improved, but an ad hoc decision cannot be easily explained.

CONCLUSION

Simulation as a discipline is like mathematics and logic. It can be studied per se to develop its own theories, methodologies, technologies, and tools; and it can be used in a multitude of problem areas in many disciplines. The uses of simulation involve this second aspect and make it a vital enabling technology for many disciplines. We have seen different aspects of simulation and lists of application areas in science, human sciences, engineering, business, finance, and defense, to name a few.

No airline would allow pilots to fly aircrafts for which they do not have relevant experience on a simulator. Similarly, proper training may be assured by using the appropriate simulator, be it for crisis management (including crisis avoidance) or emergency management training. Experiences with real-life situations are often time consuming, impractical, costly, and sometimes even deadly. Simulation avoids these types of problems for training purposes. For experimentation purposes, once a valid model is developed, simulation would allow observing the behavior of a model on different conditions including extreme and rare conditions. Several types of simulation that have been already used in hundreds of application areas are maturing and continue to be useful for experimentation and to provide experience in many conventional and challenging areas.

KEY TERMS

simulation	structural simulation	gaming simulation
experimentation	retro-simulation	predictive displays
trajectory simulation	simulation games	validation
analysis problems	constructive simulation	verification

FURTHER EXPLORATION

Simulation is used in many domains. It is rather difficult to find a domain where simulation cannot be useful. The *use* as well as *usefulness* of simulation is also evolving to cover many challenging fields. In addition to the uses of simulation listed, implied, or explained in this chapter, several tables are given in Appendix A to facilitate exploration of the activities of over 110 associations, groups, and research centers active in simulation. Their application areas would provide more specific uses of simulation.

Appendix A consists of the following tables listing simulation associations, groups, and research centers:

Table 7-A1 International (24)

Table 7-A2 By Country (32)

Table 7-A3 By Region/Language (18)

APPENDIX A   SIMULATION ASSOCIATIONS/ GROUPS/RESEARCH CENTERS

The information is updated and is also available at http://www.site.uottawa.ca/~oren/links-MS-AG.htm.

Table 7-A1 International: (24)

ABSEL—Association for Business Simulation and Experiential Learning

AMSE—Association for the Advancement of Modelling and Simulation Techniques in Enterprises

ANGILS—Alliance for New Generation Interactive Leisure and Simulation

DIGRA—Digital Games Research Association

EBEA—The Economics and Business Education Association

ESRC SAGE—Simulating Social Policy for an Ageing Society

IASTED—International Association of Science and Technology for Development

IBPSA—International Building Performance Simulation Association

IGDA—International Game Developers Association

IMA—International Microsimulation Association (a.k.a. microanalytic simulation)

IMACS—International Association for Mathematics and Computers in Simulation

INACSL—International Nursing Association for Clinical Simulation and Learning

INFORMS Simulation Society

ISAGA—International Simulation and Gaming Association (affiliated regional gaming & simulation associations can be seen at http://www.isaga.info/)

M&SPCC—Modeling and Simulation Professional Certification Commission

Modelica—Modelica Association

SAE—Human Biomechanics and Simulation Standardization Committee

SAGSET—The Society for the Advancement of Games and Simulations in Education and Training

SCS—Society for Modeling & Simulation International (formerly Society for Computer Simulation) (M&SNet, MISS)

SGI—Serious Games Initiative

SIGSIM—ACM Special Interest Group on Simulation

SSAISB—Society for the Study of Artificial Intelligence and the Simulation of Behaviour

SSH—Society for Simulation in Healthcare

WG7.1.—Modelling and Simulation Working Group of the Technical Committee TC 7 (System Modelling and Optimization) of IFIP (International Federation for Information Processing)

Table 7-A2 By Country: (30)

Australia: OzSAGA—Australian Simulation and Games Association (see http://www.isaga. info/mod/resource/view.php?id=199)

Canada: SAMS—The Society for the Advancement of Modelling and Simulation

China: CASS—Chinese Association of System Simulation

Croatia: CROSSIM—Croatian Society for Simulation Modelling

Denmark: DKSIM—Dansk Simuleringsforening (Danish Simulation Society)

Finland: FinSim—Finnish Simulation Forum

France: VerSim—Vers une théorie de la Simulation

Hungary: HSS—Hungarian Simulation Society

India: C-MMACS—Indian Society for Mathematical Modeling and Computer Simulation

Italy: ISCS—Italian Society for Computer Simulation

Italy: MIMOS (Italian Movement for Modeling and Simulation)

Japan: JASAG—Japan Association of Simulation and Gaming

Japan: JSST—Japan Society for Simulation Technology

Korea: KSS—The Korea Society for Simulation (in Korean)

Latvia: LSS—Latvian Simulation Society

Norway: NFA—Norsk Forening for Automatisering

Poland: PSCS—Polish Society for Computer Simulation (in Polish)

Romania: ROMSIM—Romanian Society for Modelling and Simulation

Singapore: SSAGSg—Society of Simulation and Gaming of Singapore

Slovenia: SLOSIM—Slovenian Society for Modelling and Simulation

Spain: AES—Spanish Simulation Society (Asociación Española de Simulación)

Sweden: MoSis—The Society for Modelling and Simulation in Sweden

UK: NAMS—National Association of Medical Simulators

UK: UKSIM—United Kingdom Simulation Society

USA: AIAA (American Institute of Aeronautics and Astronautics) M&S Technical Committee M&STC: Civil Aviation Modeling and Simulation Subcommittee, M&STC: Military Modeling and Simulation Subcommittee, M&STC: Space Modeling and Simulation Subcommittee

USA: AIMS—Advanced Initiative in Medical Simulation

USA: NASSE—The National Association of Space Simulation Educators

USA: NIA: National Institute of Aerospace—Center for Aerospace Systems Engineering, Modeling and Simulation.

USA: UrSGA—University of Rochester Simulation Gaming Association

USA: WISER—Peter M. Winter Institute for Simulation Education and Research (UPMC—University of Pittsburgh Medical Center)

Table 7-A3 By Region/Language: (18)

Asia-Pacific: APSSA—Asia-Pacific Social Simulation Association

Australia/New Zealand: MSSANZ—Modelling and Simulation Society of Australia and New Zealand

Czech and Slovak Republics: CSSS—Czech and Slovak Simulation Society

Dutch Benelux: DBSS—Dutch Benelux Simulation Society

Europe: ARGESIM (Arbeitsgemeinschaft Simulation News) Working Group Simulation News

Europe: ESSA—The European Social Simulation Association

Europe: EUROSIM—Federation of European Simulation Societies

Europe: SCS European Council

Europe: SESAM—Society in Europe Simulation Applied to Medicine

French: FRANCOSIM—Societe de Simulation Francophone

German: ASIM—German Simulation Society

Mediterranean & Latin America: IMCS—International Mediterranean and Latin American Council of Simulation

North America: NAACSOS—North American Association for Computational Social and Organizational Science

North America: NASAGA—North American Simulation and Gaming Society

North America: SAMS—Social Agent Modeling and Simulation applications section of the NAACSOS

Pacific Asia: PAAA—Pacific Asian Association for Agent-based Approach in Social Systems Research

Scandinavia: SIMS—Scandinavian Simulation Society

DKSIM (Denmark), FinSim (Finland), NFA Norway), MoSis (Sweden)

Swiss, Austrian, and German: SAGSAGA—Swiss Austrian German Simulation And Gaming Association

---

Table 7-A4 Research Centers/Groups: (25)

ACIMS—Arizona Center for Integrative Modeling and Simulation
AMSL—The Auburn Modeling and Simulation Laboratory
BioSystems Group—(UCSF—University of California San Francisco)
BMSC—Bristol Medical Simulation Centre
C&MSC—Cheshire and Merseyside Simulation Centre (Medical, Nursing, and Healthcare)
COSI—(Équipe Commande et Simulation) of LSIS, Marseille, France
CRESS—Centre for Research in Social Simulation
IBM Almaden Research Center
IST—Institute for Simulation & Training (University of Central Florida)
Liophant Simulation Club
LSIS—Laboratoire des Sciences de l'Information et des Systèmes
(Information and Systems Sciences Laboratory), Marseille, France
MSC-LES—Modeling & Simulation Center—Laboratory of Enterprise Solutions (at
  Mechanical Department, University of Calabria)
MSDS—Modelado y Simulación de Sistemas Dinámicos (Grupo Temático Español de
  CEA-IFAC), (Modeling & Simulation of Dynamic Systems (CEA-IFAC Spanish
  Thematic Group))
MSR—The Israel Center for Medical Simulation
MSREC—Georgia Tech—Modeling and Simulation Research & Education Center
MSUG—Michigan Simulation User Group
NCeSS/MoSeS—National Centre for e-Social Science/Modelling and Simulation in
  e-Social Science
NATSEM—National Centre for Social and Economic Modelling
NISAC—The National Infrastructure Simulation and Analysis Center (of the Department
  of Homeland Security's (DHS) Preparedness Directorate) SMS—Systems Modeling
  Simulation Laboratory at KAIST (Korea Advanced Institute of Science and Technology)
SRG-LBNL—Simulation Research Group at Lawrence Berkeley National Laboratory (for
  building energy simulation software)
SRG-ORMS—Simulation Research Group at Operational Research and Management
  Sciences Group at the University of Warwick
UAH CMSA—University of Alabama in Huntsville—Center for Modeling, Simulation and
  Analysis
VERSIM—VERs une théorie de SIMulation (towards a simulation theory). (VERSIM is
  part of the French research group I3 of the French CNRS (National Center of Scientific
  Research)
VMASC—Virginia Modeling, Analysis and Simulation Center

Table 7-A5 Networking of Professional Organizations: (20)

MSLS—M&S leadership Summit

SimSummit

G.A.M.E.S. Synergy Summit (Government, Academic, Military, Entertainment and Simulation)

AIMS IC—Advanced Initiative in Medical Simulation (AIMS) Industry Council (IC)

AMSC—Alabama Modeling & Simulation Council (Members & member organizations: http://resadmin.uah.edu/amsc/membersalpha2.asp)

ETSA—European Training and Simulation Association (Member organizations: http://www.etsaweb.org/index.php?module=pagemasteramp;PAGE_user_op=view_pageamp;PAGE_id=5amp;MMN_position=7:7)

IMSF—International Marine Simulator Forum (Members: http://www.imsf.org/phpBB2/memberlist.php?sid=2759ce800497dc16faf27b31159bdeb1)

ITSA—International Training and Simulation Alliance (Members: http://itsalliance.org/)

KTSA—Korea Training Systems Association

M&SNet—McLeod Modeling & Simulation Network (of SCS) (Member organizations: http://www.site.uottawa.ca/~oren/SCS_MSNet/MSNet-excom.htm)

MISS—McLeod Institute of Simulation Sciences (of SCS) (MISS centers: http://www.liophant.org/miss_world/miss_centers.html)

NCS—The National Center for Simulation (USA) (Member organizations: http://www.simulationinformation.com/cms/index.php?option=com_weblinksamp;catid=26amp;Itemid=23)

NEMSC—New England Modeling & Simulation Consortium

NIST SSC—Simulation Standards Consortium

NMASTC—National Modeling Analysis Simulation and Training Coalition

NTSA—National Training Systems Association (USA) (Membership: http://www.trainingsystems.org/membership/index.cfm)

SIAA—Simulation Industry Association of Australia (Members & member organizations: http://www.siaa.asn.au/members_current.html)

SIAA-ASSG (SISO Australia)—Simulation Industry Association of Australia Australia Standing Study Group

SUN—Simulation User Network (Medical, Nursing, and Healthcare)

UK STAG—UK Simulation and Training Action Group

Table 7A-6   Organizations/Groups/Committees (Defense): (32)

NATO (2)
NMSG—NATO Modelling and Simulation Group
SAS—NATO Studies, Analysis and Simulation Panel
By Country (30)
  Canada
    DND-CF SECO—Canadian DND/CF Synthetic Environment Coordination Office
    DND-CFEC—Canadian Forces Experimentation Centre
    DND-SMARRT—Simulation and Modeling for Acquisition Requirements, Rehearsal, and Training
  Korea
    KBSC—Korean Battle Simulation Center
  Sweden
    Swedish Defence Wargaming Centre
  Turkey
    MODSIMMER—Modeling and Simulation R&D Center
  UK
    MS&SE—Modeling, Simulation & Synthetic Environment
  USA
    DMSO—Defense Modelling and Simulation Office (US DoD)
    DRDC—Simulation and Modeling for Acquisition Rehearsal and Training
    EXCIMS—Executive Council for Modelling and Simulation (US DoD)
    MSIAC—Modelling and Simulation Information Analysis Center (US DoD)
    MSRR—Modelling and Simulation Resource Repository (US DoD)
    NDIA-M&SCom—Modelling and Simulation Committee of the Systems Engineering Division of NDIA: National Defence Industrial Association
    SISO—Simulation Interoperability Standards Organization
    SISO-SCM—SISO Standing Study Group (SSG) on Simulation Conceptual Modeling
  Air Force
    AFAMS—Air Force Agency for Modeling and Simulation
    AF-SMC/XR—Air Force SMC Modeling and Simulation Focal Point Home Page
    SDBF—Simulator Data Base Facility
  Army
    AMSO—U.S. Army Model and Simulation Office
    Battle Command, Simulation & Experimentation Directorate
    DAMO-BCM—The Army Modeling & Simulations Division
    MSRR—M&S Resource Library
    PEO STRI—U.S. Army Program Executive Office for Simulation, Training, & Instrumentation
    SMDC—U.S. Army Space and Missile Defense Command
    SNAP—Standards Nomination and Approval Process of AMSO
    TRAC—TRADOC Analysis Center
  Marine/Navy
    MTWS—MAGTF Staff Tactical Warfare Simulation
    NAVMSMO—U.S. Navy/Navy Modeling and Simulation Management Office
    NMSO—Navy Modeling and Simulation Office
  Other
    BMD SSC—Ballistic Missile Defense Simulation Support Center

REFERENCES

1. ÖREN TI (Invited Tutorial). Toward the Body of Knowledge of Modeling and Simulation (M&SBOK), In Proc. of I/ITSEC (Interservice/Industry Training, Simulation Conference). Nov. 28–Dec. 1, 2005, Orlando, FL; paper 2025, pp. 1–19.

2. ÖREN TI. The Importance of a Comprehensive and Integrative View of Modeling and Simulation. Proceedings of the Summer Simulation Conference. San Diego, CA, July 15–18, 2007.

3. ÖREN TI. Modeling and simulation: A comprehensive and integrative view. In L Yilmaz L and Ören TI, eds, *Agent-Directed Simulation and Systems Engineering*. Berlin: Wiley-Berlin; 2009 (in progress).

4. Online Etymology Dictionary (OED). http://www.etymonline.com/index.php?term=simulation. Accessed 04 March 2008.

5. Dictionary.com—Word of a Day Archive. http://dictionary.reference.com/wordoftheday/archive/2003/05/01.html. Accessed 30 March 2008.

6. BELL D. Welcome to the Post-Industrial Society. Bulletin of SCITEC (The Association of the Scientific, Engineering, and Technological Community of Canada), Oct. 1976, pp. 6–8. Reproduced from *Physics Today* Feb. 1976.

7. ÖREN TI. A Personal View on the Future of Simulation Languages (Keynote Paper). Proceedings of the 1978 UKSC Conference on Computer Simulation, IPC Science and Technology Press, Chester, England, April 4–6, 1978, pp. 294–306.

8. US House Resolution 487 (2007 July 16).

9. http://thomas.loc.gov/cgi-bin/query/z?c110:hres487. Accessed 20 March 2008.

10. http://www.stanford.edu/dept/HPS/Baudrillard/Baudrillard_Simulacra.html. Accessed 04 March 2008.

11. http://plato.stanford.edu/entries/baudrillard/. Accessed 04 March 2008.

12. ÖREN TI. Future of Modeling and Simulation: Some Development Areas. In Proceedings of the 2002 Summer Computer Simulation Conference, pp. 3–8.

13. LEVINE C. Twig, Leaf Phyllotaxy. Newsletter of the Connecticut Botanical Society, Fall 2000 (Volume 28, no. 2 & 3), http://www.ct-botanical-society.org/newsletter/phyllotaxy.html. Accessed 05 Match 2008.

14. PRUSINKIEWICZ P. Simulation Modeling of Plants and Plant Ecosystems. Communications of the ACM 2000;43(7):84–93.

15. KARPLUS W. The Spectrum of Mathematical Modelling and System Simulation. Mathematics and Computers in Simulation 1977;19:3–10.

16. JAIN S, CHAN S. Experiences with Backward Simulation based Approach for Lot Release Planning. Proceedings of the 1997 Winter Simulation Conference, pp. 773–780.

17. American Heritage Dictionary: http://www.bartleby.com/61/60/E0286000.html. Accessed 20 March 2008.

18. ÖREN TI. Bibliography of the Summer and Winter Simulation Conferences, 1967–1974. Ottawa, Ontario, Canada; 1975 (unpublished).

19. Spring Simulation Conference http://www.scs.org/confernc/springsim/springsim08/cfp/springsim08.htm. Accessed 20 March 2008.

20. SANTONI C. Contribution Méthodologique à la Conception d'Interface Homme-Machine pour les Systèmes de Supervision. Habilitation Thesis, Université Paul Cézanne Aix-Marseille III, Marseille, France. http://www.lsis/santoni/hdr/hdr.pdf. Accessed 25 March 2008. (presentation: http://www.lsis/santoni/hdr/pres-fr.ppt and http://www.lsis/santoni/hdr/hdr-eng.ppt.)

21. BALCI O. Principles and techniques of simulation validation, verification, and testing. Proceedings of the 1995 Winter Simulation Conference, Arlington, VA, pp. 147–154.

22. ÖREN TI (Invited Keynote Article). Maturing phase of the modeling and simulation discipline. In Proceedings of Asian Simulation Conference 2005 (The Sixth International Conference on System Simulation and Scientific Computing (ICSC'2005), 2005 October 24–27, Beijing, P.R.

China, International Academic Publishers–World Publishing Corporation, Beijing, P.R. China, pp. 72–85.

23. ÖREN TI (Invited Plenary Paper). Ethics as a Basis for Sustainable Civilized Behavior for Humans and Software Agents. *Acta Systemica*, 2:1, 1–5. (Also published in the Proceedings of the InterSymp 2002—The 14th International Conference on Systems Research, Informatics and Cybernetics of the IIAS, July 29–August 3, Baden-Baden, Germany.)

24. ÖREN et al. Code of Professional Ethics for Simulationists. Proceedings of the 2002 Summer Computer Simulation Conference. pp. 434–435. http://www.scs.org/ethics/scsEthicsCode.pdf. Accessed 04 March 2008.

25. Code of Professional Ethics for Simulationists. http://www.scs.org/ethics/. Accessed 30 March 2008.

Chapter 8

Modeling and Simulation: Real-World Examples

Michael D. Fontaine, David P. Cook, C. Donald Combs, John A. Sokolowski, and Catherine M. Banks

INTRODUCTION

Thus far we have discussed a number of subtopics on the subject of modeling and simulation. Part One of the text introduced the basic principles of modeling and simulation to include defining modeling and simulation, reviewing its history, and listing the advantages and disadvantages of using modeling and simulation. Part Two of our study moved the discussion to an explanation of the theoretical underpinnings of modeling and simulation that serve as the foundation for the discipline. In Part Three, *Practical Domains*, Chapter 7 introduced the many uses of modeling and simulation such as training, analysis, and decision support, to name a few. The following chapter expands the theme of practical domains by investigating real-world examples of modeling and simulation applications. The chapter is divided into four subsections: Transportation M&S, Business M&S, Medical M&S, and Social Science M&S, each written by subject matter experts. The section on Transportation M&S was written by Michael D. Fontaine, Ph.D.; Business M&S was written by David P. Cook, Ph.D.; Medical M&S was written by C. Donald Combs, Ph.D.; and Social Science M&S was written by John A. Sokolowski Ph.D., and Catherine M. Banks, Ph.D.

TRANSPORTATION MODELING AND SIMULATION

Michael D. Fontaine

The transportation system touches the day-to-day life of every member of society. A well-functioning system allows passengers and freight to reach their destination

Principles of Modeling and Simulation: A Multidisciplinary Approach, Edited by John A. Sokolowski and Catherine M. Banks.
Copyright © 2009 John Wiley & Sons, Inc.

quickly, safely, and reliably. It allows groceries to get to the corner market from distant farms, as well as allowing us to travel to work or school every day. The transportation system is inherently **multimodal**, meaning that it consists of a variety of different ways to move goods and people. Freight can be transported by air, water, truck, rail, or pipelines. Passengers travel by walking, biking, personal automobile, bus, subway, intercity rail, and airplanes. These various modes are operated by a combination of public and private sector entities and combine to create a complex system that moves goods and people throughout the world.

The state of the transportation system has become an increasing concern in the United States. Traffic growth on America's roads has far outpaced the ability of the government to build new road capacity. A study of 85 urban areas in the United States found that the rate of new road construction did not keep pace with the rate of traffic growth in any of the cities [1]. That same report estimated that the average annual delay experienced by a driver that travels in the peak hour has increased by over 270% since 1980, now averaging 38 hours per traveler per year [1]. And it's not just the roads that experience congestion.

The Federal government has documented the growing strain on the air traffic control system in the United States. In the first half of 2007, over 25% of all U.S. domestic flights arrived late, due in part to the failure to expand airport capacity to keep up with demand [2]. Port facilities are also straining to keep up. Some forecasts call for *significant* capacity problems at 75% of the nation's ports by 2010, while the total amount of port freight traffic is projected to nearly triple by 2024 [3]. As a result, the transportation community is striving to find innovative ways to deal with current and predicted congestion.

Finding solutions to address congestion on roads, ports, and in the air is often a complex problem involving multiple stakeholders and a variety of design issues. Cost effectiveness, environmental impacts, and social concerns have to be considered alongside the ability of a particular design to actually solve a problem. In many cases, the costs and timelines required to implement major transportation improvements are also significant. The Boston Central Artery/Tunnel project (commonly called the *Big Dig*) provides an example of the costs and timelines involved in a major transportation project [4]. This project involved replacing a six-lane elevated freeway with an eight- to ten-lane tunnel under downtown Boston. Design of the new 7.8-mile tunnel started in the early 1980s and the project was not substantially complete until 2006, more than 20 years after initial design started. The total cost for the project is estimated at more than $14.6 billion. Obviously, it is absolutely critical that engineers ensure that proposed solutions will actually solve the problem being examined before millions of dollars are invested and years are spent in design.

Modeling and simulation (M&S) provides an ideal tool to examine transportation problems. M&S provides a cost-effective way to assess a variety of different alternatives for addressing a problem without investing money in construction or putting the public in harm's way. M&S results then can be used to illustrate the effectiveness of different alternatives to the general public, politicians, or other key decision makers, allowing them to make informed decisions on how money should be invested.

Typical Applications of M&S in Surface Transportation

M&S is widely applied in surface transportation (roads and highways). Private consultants, cities, counties, and state departments of transportation all utilize some level of M&S to deal with common problems. Two common types of problems in surface transportation that use M&S are:

* *Transportation Planning Problems:* Transportation planning problems are focused on forecasting traffic conditions in the future, typically at least 10–20 years from the present. These analyses are often regional in scope. An example of a transportation planning problem would be predicting whether a proposed highway lane expansion would solve congestion problems 20 years from now. Macroscopic models are often used to assess these issues.

* *Transportation Operations Problems:* Transportation operations problems are usually focused on creating very detailed analyses of small scale problems, such as an individual road or intersection. An example of an operational problem would be determining how much green time should be shown to different movements at a traffic signal to reduce congestion. Microscopic models are often used to examine these problems.

A number of commercially available models are used in transportation engineering to solve these problems. Table 8-1 shows a sample of some models that are in common use. The following sections provide an introduction to how M&S is used

Table 8-1 Sample of commercially available transportation M&S software

Type of Problem	Software Name	Description
Transportation Planning	CUBE	Travel demand model
	TransCAD	GIS-based platform which includes travel demand modeling, logistics, and network analysis
	TRANSIMS	Agent based travel demand model with cellular automata model for microscopic analysis
Transportation Operations	HCS+	Deterministic, empirically based capacity analysis
	CORSIM	Microscopic, time-step traffic simulation
	VISSIM	Microscopic, time-step traffic simulation
	Paramics	Microscopic traffic simulation
	DynaMIT	Microscopic simulation designed to support intelligent transportation system evaluations
	Synchro	Deterministic traffic signal optimization
	Transyt-7F	Macroscopic simulation and traffic signal optimization
	PASSER V-03	Mesoscopic simulation and traffic signal optimization

in transportation, including an overview of important underlying concepts in these models.

Transportation Planning

One of the most common transportation planning problems involves forecasting future travel on a roadway network. Virtually every urban area maintains a macroscopic **travel demand model** that is used to generate these forecasts. The projections generated by the models are then used by local politicians to prioritize where improvements should be made to the road or public transportation system. These models use socioeconomic data and roadway information to predict conditions on the roadway network in future years. These models are typically macroscopic in nature, although several new models are trying to deal with this problem in a more microscopic way.

Four-Step Travel Demand Models

Historically, travel demand models have used an approach that is generically called the **four-step model**. The four step model is a macroscopic approach that attempts to predict future travel on a network over some long time horizon, typically 20 years or more. The four steps used in this approach are:

- trip generation
- trip distribution
- mode split
- traffic assignment

The basic inputs to the model are forecasts of future land use, employment, and population throughout a region, as well as information about the projected future transportation network. People make a trip to serve a particular purpose, like driving to work or the store. The model uses socioeconomic data to estimate how much travel is going to occur as a result of people carrying out these day-to-day tasks in a future year.

In these models, socioeconomic data is aggregated into homogeneous areas termed **traffic analysis zones** (TAZs). TAZs can vary in size from just a few blocks in an urban area to up to five square miles in more rural locations. Travel between TAZs is forecasted based on the socioeconomic data in each zone. The main outputs of these models include the forecasted traffic volume on the roads, the speed on roads, and emissions estimates. The following sections briefly describe each of the four steps in this type of travel demand model. Some numerical examples of key calculations are presented. More detailed discussion of these models can be found in the text by Ortuzar and Willumsen [5].

Trip Generation   A trip is defined as a one-way movement from an origin to a destination. In the trip generation step, the number of trips that are expected to start or end in each TAZ are estimated. Trips might occur by any means, such as car, walking, biking, or bus. Essentially, this estimates the number of trips that will start or end in a TAZ based on socioeconomic and land use factors. The number of trip productions (number of trips that start in a TAZ) is often influenced by income (higher income households have more trips), household characteristics (larger households generate more trips), and car ownership (more cars generate more trips). Trip attractions (number of trips that end in a TAZ) are often related to factors like the amount of square footage of retail/commercial space, the number of parking spaces, and the projected employment in an area.

The estimated number of trips can be determined several ways. Growth factors can be applied to existing estimates of trip productions and attractions. Growth factor models typically assume a growth factor that is a function of population, income, and car ownership. The growth factor approach is relatively crude, since it essentially assumes a linear extrapolation based on current conditions. Regression models are more commonly used to perform trip generation since they can directly predict the number of trips as a function of a variety of socioeconomic factors. For example, trip productions may be directly predicted as a function of population, income, auto ownership, and age of household members. Trip attractions in business districts may be predicted as a function of square foot of retail/commercial space and number of employees. Cross-classification techniques can also be used to estimate trip generation rates for a specific set of socioeconomic variables.

Trip Distribution   In trip generation, the number of trips expected to start or end in a particular zone is estimated. In trip distribution, the goal is to estimate where the trips that start in a TAZ are going and where the trips that end in a TAZ are coming from. One common model used in the step is called the gravity model. The gravity model is defined as:

$$T_{ij} = P_i \frac{A_j f(L_{ij})}{\sum_{j'} A_j f(L_{ij})} \qquad (8.1)$$

where:

T_{ij} = number of trips starting in TAZ i and traveling to TAZ j

P_i = number of trips starting in TAZ i (Productions)

A_j = number of trips ending in TAZ j (Attractions)

$f(L_{ij})$ = impedance function for the path between TAZ i and TAZ j

The **impedance function** is typically based on the travel time or cost to go between two TAZs. Longer travel times and higher costs discourage trips. The net result of the gravity model is that more trips will be predicted between closer destinations than between far away destinations.

A key issue with gravity models is that the number of productions must equal the number of attractions. This is termed flow conservation. The gravity model is underspecified, which means that there are an infinite number of potential solutions. As a result, a structure is required to solve these problems and corrections must be made to the values generated by equation 8.1. The following example helps illustrate how the gravity model is applied in the trip distribution step.

Example: Gravity Model

Consider the four TAZs shown in Table 8-2. The trips produced by and attracted to each zone are shown in the table. At this point, the number of trips that start in a TAZ (productions) and the number that end in a TAZ (attractions) are known from the trip generation step. We don't know where the productions are going or where the attractions are coming from, so we need to determine how the TAZs relate to one another. Note that the sum of the productions is equal to the sum of the attractions (4350 trips).

Table 8-3 shows the results of the impedance function between each set of zone pairs. This is typically a function of travel time and/or cost to travel between zones. A detailed discussion of impedance functions can be found in Ortuzar and Willumsen [5].

The first step in applying the gravity model is to develop an unconstrained trip matrix. This is done by multiplying the impedance function results from Table 8-3 by the number of productions and attractions for each production/attraction zone combination—$P_iA_jf(L_{ij})$. For example, if we wanted to find the unconstrained number of trips between production zone 1 and attraction zone 2, the value would be:

$$(\text{Productions in Zone 1}) \times (\text{Attractions to Zone 2})$$
$$\times (\text{Zone 1 to 2 Impedance Function}) = 900(1{,}325)(4) = 3{,}040{,}000.$$

Table 8-2 Productions and Attractions by TAZ

TAZ	Productions	Attractions
1	900	800
2	950	1,325
3	1,300	900
4	1,200	1,325
Total	4,350	4,350

Table 8-3 Impedance Values Between TAZs

Production Zone	Attraction Zone			
	1	2	3	4
1	20	4	12	4
2	4	20	4	12
3	12	4	9	2
4	4	12	20	9

Table 8-4   Unconstrained Trip Matrix

Production Zone	Attraction Zone			
	1	2	3	4
1	14,400,000	4,770,000	9,720,000	4,770,000
2	3,040,000	25,175,000	3,420,000	15,105,000
3	12,480,000	6,890,000	10,530,000	34,450,000
4	3,840,000	19,080,000	21,600,000	14,310,000

Table 8-5   Scaled Trip Matrix Showing Ratios to Target Values

Production Zone	Attraction Zone				Total P_j	Target	Ratio
	1	2	3	4			
1	307.7	101.9	207.7	101.9	719.2	900	1.251
2	65.0	537.9	73.1	322.8	998.7	950	0.951
3	266.7	147.2	225.0	736.1	1,375.0	1,300	0.945
4	82.1	407.7	461.5	305.8	1,257.1	1,200	0.955
Total A_i	721.5	1,194.7	967.3	1,466.6	4,350		
Target	800	1,325	900	1,325			
Ratio	1.109	1.109	0.930	0.903			

The total of all of the trips in Table 8-4 is 203,580,000, which is substantially different from 4,350 trips shown in Table 8-2. If we divide the total number of unconstrained number of trips by 4,350, we get a factor of 46,800. Next, we divide each of the cells in Table 8-4 by 46,800 to normalize the trip matrix, resulting in the scaled trip matrix shown in Table 8-5. Table 8-5 has the correct total number of trips, but the productions and attractions are still not equal to what was in the Table 8-2 for each zone. Next, we need to compare the scaled trip matrix to the target production and attraction values and determine the ratios between the target number of trips and the current number. Table 8-5 shows the ratio of productions and attractions to the target values given Table 8-2.

For our first iteration, we will multiply each of the trip totals by the ratio for the appropriate production zone. As an example, each cell for production zone 1 in Table 8-5 will be multiplied by the ratio of 1.251. This process is repeated using the appropriate ratio for each of the other production zones. This generates the trip matrix shown in Table 8-6.

In Table 8-6, the productions are equal to the targets but the attractions are not adding up to the appropriate target number. Next, we multiply the ratios for attractions by the columns in the trip interchange matrix and recalculate the number of trips. This produces Table 8-7.

Table 8-6   Trips Scaled by Production Ratios

Production Zone	Attraction Zone				Total P_j	Target	Ratio
	1	2	3	4			
1	385.0	127.5	259.9	127.5	900	900	1.0
2	61.8	511.7	69.5	307.0	950	950	1.0
3	252.1	139.2	212.7	696.0	1,300	1,300	1.0
4	78.3	389.2	440.6	291.9	1,200	1,200	1.0
Total A_i	777.3	1,167.7	982.7	1,422.4	4,350		
Target	800	1,325	900	1,325			
Ratio	1.029	1.135	0.916	0.932			

Table 8-7   Trips Re-scaled by Attraction Ratios

Production Zone	Attraction Zone				Total P_j	Target	Ratio
	1	2	3	4			
1	396.3	114.7	238.0	118.8	897.9	900	1.002
2	63.6	580.7	63.7	285.9	993.9	950	0.956
3	259.5	158.0	194.8	648.3	1,260.6	1,300	1.031
4	80.6	441.6	403.5	271.9	1,197.6	1,200	1.002
Total A_i	800	1,325	900	1,325			
Target	800	1,325	900	1,325			
Ratio	1	1	1	1			

Table 8-7 shows that the differences between our target values and our actual production and attraction values are growing smaller. This process needs to be repeated for multiple iterations, alternating between applying the production and attraction ratios until the trip interchange matrix converges to a solution that satisfies the required number of productions and attractions by zone. Table 8-8 shows the final trip interchange matrix, generated after several more iterations. This example provided a relatively simple case. A metropolitan area could easily have hundreds of zones, creating a problem that is much more difficult to solve.

While gravity models are commonly used, they also have some drawbacks. One of the problems is that all TAZ pairs with the same impedance function values are treated the same. There is no direct consideration of socioeconomic characteristics that might influence travel. As a result, it is possible that the results of the gravity model may not perfectly match reality.

Mode Split   In the mode split step, the proportion of traffic traveling by different modes of transportation (such as car, walking, and public transportation) is estimated. The choice of mode is influenced by a variety of factors, including the characteristics

Table 8-8 Final Trip Interchange Matrix

Production Zone	Attraction Zone				Total P$_j$	Target	Ratio
	1	2	3	4			
1	394.7	148.3	238.2	118.8	900.0	900	1.0
2	59.7	560.8	60.1	269.5	950.0	950	1.0
3	265.9	166.5	200.7	666.9	1,300.0	1,300	1.0
4	79.7	449.4	401.0	269.9	1,200.0	1,200	1.0
Total A$_i$	800.0	1,325.0	900.0	1,325.0			
Target	800.0	1,325.0	900.0	1,325.0			
Ratio	1.0	1.0	1.0	1.0			

of the person making the trip, the reason for making the trip, and the characteristics of the modes available. Logit models are one option for predicting mode split. After the number of trips between TAZs is known, the logit models can be used to predict the proportion of people using different modes. The simplest form of logit model is a binary model that computes the percentage of people using one of two available modes. For example, a model then calculates the share of people using bus versus car may look like:

$$S_B = \frac{e^{U_B}}{e^{U_A} + e^{U_B}} = \frac{1}{1 + e^{(U_A - U_B)}} \tag{8.2}$$

where:

S_B = % of people using bus

U_B = utility of bus mode

U_A = utility of automobile mode

The utility function is developed from regression models that show how people have made travel decisions as a function of factors like travel time and cost. The utility functions tend to be site-specific since availability of modes and alternate routes plays an important role in mode choice decisions.

More disaggregate methods are also available that subdivide mode choice into different socioeconomic groups. This is often more accurate since certain groups may be more likely to use certain modes. For example, low income groups are more likely to be *captive* transit riders who have limited opportunity to travel by personal automobile.

Traffic Assignment In the traffic assignment step, vehicles are assigned to specific routes to travel between their intended origins and destinations. This is done until each vehicle has been assigned to the route that minimizes its individual travel time between its origin and destination. This is called user equilibrium assignment. In this form of assignment, no one can unilaterally change routes and improve their travel time.

There are several algorithms that can be used to assign vehicles to specific routes. One method is called the **Frank-Wolfe algorithm**. The Frank-Wolfe algorithm essentially uses a descent approach to minimizing an objective function. In this method, functions defining the travel time on the network are defined, and then traffic flow is iteratively assigned to links until a user equilibrium state is achieved. Assignment is usually stopped after a set number of iterations or when link flows change by some minimal amount. Ortuzar and Willumsen provide several examples of the application of the Frank-Wolfe algorithm and other options for performing traffic assignment [5].

TRANSIMS

One of the limitations of the four-step travel demand model is that it is inherently macroscopic and does not contain detailed consideration of the impact of individual traveler characteristics on travel choices. The *Transportation Analysis and Simulation System* (**TRANSIMS**) is an alternative approach to the four-step model that deals with travel in a much more disaggregate manner [6]. It uses an agent-based approach and microscopic simulation to consider an individual's daily activities. This allows for the explicit consideration of what is termed a **trip chain**—allowing the model to examine a trip starting at a home, going to the grocery store, then the dry cleaners, then a gas station, and then home again. This is in contrast to the four-step model that only considers the origin and destination of a trip, and not the linkages between what someone would actually do in their daily life. In theory, TRANSIMS should produce more accurate and detailed information about travel than what could be derived from the four-step model. The basic operation of TRANSIMS can be summarized as follows:

- First, census data is used to create synthetic households for the study area. Each individual person in the area is assigned to a household with distinct characteristics based on the aggregate distributions in the census data. In effect, this creates simulated individual households with distinct characteristics, as opposed to the TAZ level analysis performed in four-step travel demand models.

- Next, activities are generated for each household based on the characteristics of the individuals. This list of activities is what will dictate the person's travel patterns throughout the day. This data is typically derived from household travel surveys that many planning agencies conduct regularly.

- Routes are assigned for each trip generated in the second step. This is more detailed than what you would normally see in a four-step model, including consideration of travel mode, parking locations, and even who the person is traveling with.

- A microscopic traffic simulation is then performed using the route plans developed in the previous step. In this case, a cellular automata model is used to estimate traffic conditions at a microscopic level. This represents a significant increase in the fidelity of the model over what is typically generated in a four-step model. More detailed data on link performance is included, and traveler experiences can be looked at in a more fine-grained manner than with the four-step model. Furthermore, the influence of stochastic changes in traffic can be estimated, allowing the analyst to develop confidence intervals for performance. This is not possible in a four-step model.

This process is iterative, since travel decisions and route choice are directly affected by the performance of the transportation system. The results of the microscopic traffic simulation will show which routes are congested at certain times of the day. This information is fed back into the activity generation and route choice modules since it could directly impact the decision to make a trip as well as the route selected. This process is completed until convergence is reached.

While TRANSIMS offers a number of conceptual benefits, it has yet to be widely adopted by practitioners. One of the most significant barriers to the widespread adoption of TRANSIMS is the large data requirements to build the model. Significantly more information is required than with the four-step model since the movements of individual travelers are simulated over a 24-hour period. For example, the microscopic traffic simulation requires significant amounts of data that may not be readily available, such as detailed data on the length of right and left turn bays and traffic signal timings. This data is not needed in most four-step models since traffic is simulated on a macroscopic level.

Transportation Operations

Transportation operations problems are usually focused on examining relatively small scale problems related to congestion and traffic flow. In this case, engineers are focused on examining specific issues related to how a road is designed or controlled, rather than trying to forecast regional trends in travel. As a result, operational models often use microscopic methods to examine how individual drivers might react to specific design elements. Traffic signal operations, freeway merging and weaving, and access to property are examples of operational issues that would be assessed using these methods. There are several distinct ways to perform an operational analysis, ranging from macroscopic deterministic methods to sophisticated microscopic models.

Highway Capacity Manual Deterministic Models

Originally published in 1950, the *Highway Capacity Manual* (HCM) represents one of the most well-established references in traffic engineering [7]. The HCM defines a set of empirically derived, deterministic models that are used to describe operational performance for everything from freeways to intersections with traffic signals. The term **capacity analysis** is used to describe analyses performed using HCM procedures that quantitatively determine the ability of a road to process traffic. The term *capacity* refers to the maximum number of vehicles that can pass a point on a road in a certain period of time. Until the widespread availability of microscopic traffic simulation models in the 1980s, the HCM represented the only way to perform an operational analysis. HCM procedures are now implemented in several widely available computer models.

The HCM provides unique models for each type of road facility. For example, empirical equations are developed to predict the amount of delay drivers experience at intersections with traffic signals or stop signs, and other models predict speed and traffic density for freeway sections. A common theme throughout all HCM models is the concept of **level of service** (LOS). The HCM grades the performance of a particular facility (the level of service) on an "A" to "F" scale depending on the model results. The LOS concept is used to assist in communicating results with politicians and other decision makers who may not have an idea of what acceptable performance is on a particular section of road. HCM capacity analysis has developed into a de facto standard in traffic engineering, although many agencies are moving increasingly towards simulation-based analyses. The LOS concept in heavily ingrained into the transportation culture in the United States, and many transportation agencies require an HCM analysis even if more sophisticated simulation models are used.

The widespread acceptance of HCM techniques is certainly a major advantage of these models, but they also have a number of limitations. First, the HCM models function best when looking at isolated intersections or road segments. They do not generally do a good job of examining complex networks where different roadway elements interact with one another. The HCM is also deterministic, which is not an accurate way to model traffic. Traffic by its very nature is stochastic, with driver behavior and traffic volume rates varying over time. Finally, the concept of LOS does not reflect variations in what is considered acceptable performance from region to region. For example, a LOS of "D" is considered unacceptable for many rural signalized intersections, while it would be considered acceptable in an urban area where drivers have a higher tolerance for delay.

Microscopic Simulation Models

A number of **microscopic traffic simulation** models have been developed to analyze the operational performance of roads. These models provide extremely detailed information on what delays, travel speeds, and traffic backups might be encountered by drivers on a road. The performance of individual vehicles is simulated, enabling detailed analyses of critical traffic control elements like traffic signals. Some models can also simulate pedestrians, bicyclists, buses, and light rail systems as well. Simulation results are generally superior to HCM model results if multiple roadway elements (such as traffic signals or freeway ramps) interact with one another. They also provide tools that allow for the visualization of traffic flow through the network. Figure 8-1 shows an example of a visualization produced by a microscopic traffic simulation.

Figure 8-1   Example of microscopic traffic simulation visualization

Some of the most widely used models are stochastic, time step–based models that simulate individual vehicles traveling through a roadway network. Driver characteristics such as gap acceptance, desired speed, and braking/acceleration aggressiveness are generated for each vehicle that enters the network. Using simulation allows the random variation of the real world to be accounted for when traffic studies are performed.

Traffic simulation models have several typical data requirements. First, the user must know the basic layout of the road system being examined (number of lanes, locations of intersections, etc). The user must also know the traffic control present on the roads. This includes the speed limits on the road, locations of stop and yield signs, and the location and timing of traffic signals. Basic elements of driver behavior should also be known, such as driver compliance with speed limits and how quickly vehicles start moving when a traffic signal changes from red to green. Finally, the user needs to know about the traffic on each roadway link. The number of vehicles, the type of vehicle (car/truck/bicycle, etc.), and how many vehicles make a particular turn at an intersection are commonly input into the model. Data requirements for these models are usually fairly significant, and field data collection is often required to ensure that accurate input data is used.

Car Following Theory and Lane Changing Logic

Car following theory describes how drivers accelerate and decelerate based on the traffic that is around them. This forms the underlying basis for the traffic dynamics within a simulation model. A number of researchers have developed equations to represent car following behavior [8]. One relatively simple car following model was developed by researchers at General Motors (GM):

$$a_{n+1}(t + \Delta t) = \frac{\alpha_0}{x_n(t) - x_{n+1}(t)} [v_n(t) - v_{n+1}(t)] \qquad (8.3)$$

where:

n = lead vehicle

n + 1 = following vehicle

t = time (s)

Δt = reaction time of driver (s)

x = position of vehicle at a particular time relative to an origin (ft)

v = speed of vehicle (ft/s)

a = acceleration or deceleration rate of a vehicle (ft/s^2)

α_0 = sensitivity constant (ft/s)

The GM equation predicts the acceleration of a following vehicle based on the relative distance to the vehicle it is following and the difference in the speeds of the two vehicles. There are several interesting features to this equation. First, it can be seen that the reaction time of the driver is incorporated into the model. The driver of the following vehicle is deciding whether to speed up or slow down at time t + Δt based on the vehicle characteristics at time t. As a result, if the leading vehicle speeds up or slows down, then there is a lag time Δt until the following vehicle reacts. This equation also includes sensitivity to the distance between the leading and following vehicles. Long distances between the vehicles will generate more gradual acceleration/deceleration profiles, while short distances will tend to create more rapid changes in speed as drivers seek to avoid collisions. Equations like this are used to define how vehicles interact in a microscopic traffic simulation.

Example: Car Following Models

A simple example will help to illustrate how a car following model is applied. Consider a case where two cars are at a complete stop at a traffic signal, with the following car 25 feet behind the leading car. Starting at time t = 0, the signal turns green and the leading car begins to accelerate at a rate of 4 ft/s^2. It maintains this acceleration for 10 seconds, and then begins to decelerate at a rate −2 ft/s^2 for another five seconds. A reaction time (Δt) of 1 second and a sensitivity constant (α_0) of 30 ft/s are assumed. The goal is to track the speed and position of both vehicles every second for the 15-second period. Assume that accelerations of the following vehicle are only updated at discrete time steps equal to whole seconds, and they are maintained until the next update.

In this example, basic kinematic equations from physics are used to first determine the speed and position of the leading vehicle assuming constant acceleration/deceleration:

$$x_n(t + \Delta t) = x_n(t) + v_n(t) \times (\Delta t) + \frac{a_n(t) \times (\Delta t)^2}{2}$$

$$v_n(t + \Delta t) = v_n(t) + a_n(t) \times (\Delta t)$$

Using these equations, at time t = 1 sec:

$$x_{leader}(t=1) = x_{leader}(t=0) + v_{leader}(t=0) \times (1) + \frac{a_{leader}(t=0) \times (1)^2}{2}$$

$$= 0 + 0 \times 1 + \frac{4 \times (1)^2}{2} = 2 \, \text{ft}$$

$$v_{leader}(t=1) = v_{leader}(t=0) + a_{leader}(t=0) \times (1) = 0 + 4 \times (1) = 4 \, \text{ft/s}$$

At time t = 2 sec:

$$x_{leader}(t=2) = x_{leader}(t=1) + v_{leader}(t=1) \times (1) + \frac{a_{leader}(t=1) \times (1)^2}{2}$$

$$= 2 + 4 \times 1 + \frac{4 \times (1)^2}{2} = 8 \, \text{ft}$$

$$v_{leader}(t=2) = v_{leader}(t=1) + a_{leader}(t=1) \times (1) = 4 + 4 \times (1) = 8 \, \text{ft/s}$$

This process is repeated until time t = 15 seconds is reached. Next, the acceleration of the following vehicle has to be calculated. Since there is a lag of Δt = 1 second and the leading vehicle starts from a full stop, the trailing vehicle would not begin to accelerate until t = 2 seconds. At time t = 2 seconds:

$$a_{follower}(t=2) = \frac{\alpha_0}{x_{leader}(t=1) - x_{follower}(t=1)} [v_{leader}(t=1) - v_{follower}(t=1)]$$

$$a_{follower}(t=2) = \frac{30}{2-(-25)} [(4)-(0)] = \frac{30}{27}(4) = 4.44 \, \text{ft/s}^2$$

Speeds and distances for the following vehicle would then be updated using the kinematic equations, just like the leading vehicle. At time t = 3 seconds:

$$a_{follower}(t=3) = \frac{\alpha_0}{x_{leader}(t=2) - x_{follower}(t=2)} [v_{leader}(t=2) - v_{follower}(t=2)]$$

$$a_{follower}(t=3) = \frac{30}{8-(-25)} [(8)-(0)] = \frac{30}{33}(8) = 7.27 \, \text{ft/s}^2$$

Following another update of speeds and velocities for the following vehicle, at time t = 4 seconds:

$$a_{follower}(t=4) = \frac{\alpha_0}{x_{leader}(t=3) - x_{follower}(t=3)} [v_{leader}(t=3) - v_{follower}(t=3)]$$

$$a_{follower}(t=4) = \frac{30}{18-(-22.78)} [(12)-(4.44)] = \frac{30}{40.78}(7.56) = 5.56 \, \text{ft/s}^2$$

Table 8-9 summarizes the speed, acceleration, and position of the leading and following vehicle for the 15 seconds being analyzed. As can be seen in Table 8-9, there is a delay between when the leading and trailing vehicles begin to decelerate. In a microscopic traffic simulation, similar equations would govern the interactions between all vehicles in the network. This example can

Table 8-9   Example Application of General Motors Car Following Equation

Time (sec)	Lead Vehicle			Following Vehicle		
	Acceleration (ft/s^2)	Speed (ft/s)	Distance (ft)	Acceleration (ft/s^2)	Speed (ft/s)	Distance (ft)
0		0	0		0	−25.00
	4			0		
1		4	2		0	−25.00
	4			0		
2		8	8		0	−25.00
	4			4.44		
3		12	18		4.44	−22.78
	4			7.27		
4		16	32		11.72	−14.70
	4			5.56		
5		20	50		17.28	−0.20
	4			2.75		
6		24	72		20.03	18.45
	4			1.63		
7		28	98		21.66	39.29
	4			2.23		
8		32	128		23.88	62.06
	4			3.24		
9		36	162		27.12	87.56
	4			3.69		
10		40	200		30.82	116.53
	−2			3.58		
11		38	239		34.39	149.14
	−2			3.30		
12		36	276		37.70	185.18
	−2			1.20		
13		34	311		38.90	223.48
	−2			−0.56		
14		32	344		38.34	262.10
	−2			−1.68		
15		30	375		36.66	299.60

be simulated using the continuous simulation spreadsheet technique discussed in Chapter 3.

A second important component of microscopic traffic simulations is lane changing logic [9]. In microscopic models, drivers must make a decision about when it is safe to change lanes or make turns. This is typically done based on the available gaps in traffic in the traffic lane that a vehicle is merging into or crossing through. A gap is defined as the space between the back bumper of a leading car and the front bumper of the car following behind it. A gap can be defined in terms of time or distance. Each vehicle is assigned a critical gap that defines the minimum gap that a driver will accept to make a lane change or turn. If the available gap exceeds the critical gap, the maneuver will be executed. If not, the driver will wait for another gap. Lane changing logic also typically includes a consideration of the vehicle's speed, allowing a vehicle to accelerate or decelerate to change lanes.

Microscopic models typically separate lane changing maneuvers into three categories. Mandatory lane changes occur when a vehicle is forced to move from its current lane. For example, a mandatory lane change might occur when the lane the vehicle is traveling in ends. A discretionary lane change occurs when a driver wants to change lanes because they are not traveling at their desired speed. This would include cases where a driver is caught behind a slow moving vehicle and wants to pass so they can travel at a faster speed. Finally, an anticipatory lane change involves a vehicle changing lanes to avoid congestion ahead or to ensure that the vehicle is in the proper lane to make a downstream turn. The critical gaps for these maneuvers would vary based on the type of lane change.

Model Calibration

Calibration is extremely important in traffic simulations. Driver behavior can vary significantly from region-to-region, and may even change significantly within a city. For example, drivers in congested, urban areas tend to be willing to accept smaller gaps in traffic when making turns than those in more rural, uncongested areas. Using model default values may result in poor agreement between the simulation results and reality. A three-step strategy for calibration of microscopic traffic simulation models has been suggested [9]:

- *Capacity Calibration.* If a road is expected to operate near its maximum capacity, it is important to ensure that the simulation is accurately reproducing what is seen in the real world. For example, a typical lane of freeway can carry a maximum of 2000 to 2400 vehicles per hour. Simulation parameters that have the potential to impact capacity should be investigated to ensure that there is a good match between the simulation and real data. In the absence of real data, HCM estimates of capacity are sometimes used.

- *Route Choice Calibration.* The traffic entering and exiting the model at key locations should be examined to ensure that there is a good match between what was observed in the field and what is produced by the model.

- *System Performance Calibration.* Finally, field measurements of certain performance measures (such as travel time, queue length, or delay) should be compared to what was generated by the model. If the simulation results are found to differ, driver behavior parameters are then altered until a good match is achieved with the field data.

System performance calibration often represents the most significant effort in the calibration process. It is very important that appropriate calibration measures be selected, since field data collection is usually involved. Measures should be relatively easy to collect, be sensitive to changes in traffic, and the model should calculate the measure in the same way as it is collected in the field [10]. At an intersection with a traffic signal, the travel time through the intersection continues to increase as traffic volumes increase (a good calibration measure). Once the intersection reaches capacity, the total number of vehicles that go through the intersection will level off since the signals will meter the flow (a bad calibration measure).

Common system performance calibration measures used are delay, travel time, vehicle counts, speeds, and queue lengths. Initially, the simulation model is run with the default driver behavior values and the field data is compared to the same measures generated by the simulation [9]. Certain factors related to the car following model are then changed until there is a good agreement between the simulation model and field data. This might involve changing factors such as average following distance, driver reaction time, critical gaps, traffic signal startup lost time, and queue discharge headways [10]. In some cases, these factors (such as startup lost time and critical gap) can be measured in the field. Other factors may need to be examined through an iterative search process or other technique.

This discussion on Transportation M&S focused primarily on how M&S is applied in the area of surface transportation (roads and highways) where a rich body of literature exists that illustrates the range of transportation models and applications. M&S is also an important tool in examining other modes of transportation. Students interested in these and other aspects of the transportation challenge are encouraged to explore this subject—it is undoubtedly a challenge that we must address as we progress into the 21st century.

BUSINESS MODELING AND SIMULATION

David P. Cook

Industrial organizations have long employed simulation as a decision support methodology. Though the date of the first business simulation model developed for use in a private company is not known, the first private industrial organizations began to apply simulation as a decision support tool in the late 1940s. Today, simulation models are widely deployed across business organizations. The near-ubiquity of simulation models can be attributed in part to the business results that can be achieved through the use of simulation. Advances in hardware and software technologies have made simulation accessible. Today, for example, many managers have at their fingertips powerful simulation tools embedded in spreadsheet applications in order to rapidly conduct what-if analyses. Simulation has migrated from the mainframe to the desktop in a relatively short time-frame. According to a mid-1990s survey, simulation was employed at 90% of responding organizations [11]. Simulation is widely deployed across a broad spectrum of business areas including production/operations, planning/corporate strategy, marketing/sales, personnel/human resources, accounting, and finance.

The traditional focus of business-oriented simulations has been in manufacturing; however, in recent years simulation has been increasingly applied to the service sector [12]. For example, simulation models of service call centers, health care systems, and air transportation systems are regularly developed [13–15]. The rise in service operations simulations should be no surprise since services represent a significant share of economic activity in developed and developing countries around the world.

The application of simulation to business and business problems can be categorized in a variety of ways. For example, the categorization of real-world business applications of simulation can be based on the objectives of the simulation. The typical business objectives of simulation can be categorized as system analysis, acquisition and system acceptance, research, and education and training. In **system analysis**, a system analog is created for the purpose of gaining an understanding of the system's behavior or to improve its performance. In the context of a business simulation study, the system of interest may include labor, capital, or labor and capital. **Acquisition and system acceptance** simulation studies generally involve answering questions related to whether a system meets minimum performance requirements or a subsystem is capable of improving the performance of the larger system of which it is a part. Research often involves the development of an artificial environment. In the context of this hypothetical environment, the model developer is able to test the effects of changes to system components, or the behaviors of groups and/or individuals on system performance. Education and training simulations are intended to develop, in students and trainees, a greater understanding of and ability to apply business ideas, concepts, and techniques [16].

Describing simulations on the basis of how time and state are handled such as Monte Carlo simulation or discrete-event simulation or continuous simulation, is also a reasonable basis on which to categorize simulations applied to business settings. In **Monte Carlo simulation** models time is of little consequence; however, the ordering of events is a focal point. In discrete-event simulations, state variables change at discrete points in time. Continuous simulations, on the other hand, have state variables that change continuously over time.

The following discussion is a representative sample of simulations applied to business. The purpose is not to cover every business application; nor is it to present a complete sample of simulation types. The intent is to introduce you to the extent of the variety of simulation for the purpose of education and training and in such functional areas as production and service operations, management, marketing, and finance.

Business Applications

Training and Education

The focus of this chapter is real world examples for the application of modeling and simulation. The inclusion of simulation for the purpose of training and education, especially those training and education simulations reliant upon gaming, may seem out of place in a chapter such as this. However, the use of simulation in business for the purposes of training and education has a rich history stretching back more than 50 years. The first widely known business game, **Top Management Decision Simulation**, was developed by the American Management Association in 1956 for use in some of its management seminars [17]. Since that time, business training and education simulations have grown in popularity and sophistication.

Many sophisticated and challenging business issues are difficult to effectively convey to students/trainees, in the context of a purely hypothetical setting. They need to have some way to more concretely experience such issues [18]. Business simulation games provide a relatively low-cost method to give students this meaningful experience. Participants can make errors in a simulated environment that could cost many thousands of dollars in a live environment. University students who have this type of experience can be productive more quickly than their contemporaries. In the early 1990s, a manufacturing planning and control simulation called **ITEC** was developed to provide students with experience in managing materials flow through complex manufacturing facilities in order to gain a greater understanding of the strategic and tactical decisions that are required to coordinate a manufacturing plant's production activities. A professor commenting on the experience of a recent graduate who had participated in the simulation exercise indicated she was told by her new employer that she was months ahead of other recent graduates due to her understanding of the manufacturing planning and control activities developed through the ITEC simulation [19].

Manufacturers have deployed sophisticated simulations to train employees and upgrade their skills. RealVue Simulation Technologies developed software to simulate complex equipment in order to allow service technicians, machine operators, and others to train on the maintenance, repair, and operation of equipment without having to be physically at the equipment itself. This can reduce, or eliminate, the need to purchase equipment for the sole purpose of training. This helps to reduce the opportunity costs associated with occupying equipment for the purpose of training as opposed to production and the cost associated with inexperienced personnel damaging equipment during training cycles. Numerous advantages of such training technologies have been cited: simulation training software can help to reduce training time and costs; training can occur at the trainee's learning pace; the simulation can be designed to allow students to make mistakes; and the simulation can be used as on-the-job help [20].

Real-World Business Applications

The examples presented here are condensed versions of research reports. The purpose is not to provide an intense examination of the problems contained in the cited research reports, but to give you a sense of the nature of the problems that commonly face business managers. It is also important for you to recognize that simulation is oftentimes a complementary problem solution methodology. It is frequently used in conjunction with advanced statistical analysis tools and the traditional tools of management science.

Business simulation can be used as an agent for strategic change. Two examples of the application of simulation for this purpose are presented. In the first case, traditional management science techniques in conjunction with simulation are used to help reshape operations strategy at the Standard Register Company [21]. In the second example, a simulation is used to reconfigure a remanufacturing line at a Visteon facility in Mexico in order to alleviate capacity shortage problems [22]. In both cases, simulation is used to develop solutions to real-life business problems common in industry.

Standard Register Company Standard Register Company is a leading document services company with nearly a century of experience in the industry.[1] It offers traditional print as well as digital information-capture and management services to the healthcare, financial services, manufacturing, and other industries. Standard Register is an established leader in high-volume print production of forms and stationary for major U.S. firms. Healthcare bill statements, financial statements, and retail target-marketing mailers are examples of traditional, high-volume, low-cost print jobs. Such jobs, the *rotary* product, are typically run on continuous-feed rollers and subsequently cut, folded, and trimmed to customer specifications.

A shift in the document services industry to new product lines involving digital formats has eroded the market for traditional print documents. Standard Register, as well as other companies competing in the market for printed business documents, is being faced with oversupply and stiff price competition. To remain competitive in this highly competitive business, Standard Register must make tough operational decisions in order to offer competitive pricing to its customers.

To be in a position to be price-competitive, Standard Register must operate more cost efficiently. A study was designed to facilitate the optimal allocation of production orders across its production-distribution network in order to minimize total landed cost. Total landed cost is the sum of all the costs associated with making and delivering the product to the point where they produce revenue, usually the customer. This includes not only production costs, but also transportation costs. For a company such as Standard Register producing heavy paper products,

---

[1] Adapted from Ahire SL, Gorman MF, Dwiggins D, Mudry O. "Operations research helps reshape operations strategy at Standard Register Company." *Interfaces* 2007;37(6):553–565.

transportation costs are a significant cost factor. Simply being capable of producing a product at minimum cost is insufficient if the point of production is far from the customer.

The production process associated with printing forms at one of Standard Register's plants can be divided into three phases: pre-print operations, printing operations, and post-printing operations. Based upon product specifications (e.g., roll or cut sheet, number of colors, finishing options) jobs are scheduled for completion at a specific printing press at a particular plant. Standard Register operates a geographically dispersed network of 13 facilities consisting of approximately 150 printing presses, each with different capabilities and efficiencies. Some jobs cannot be completed on certain presses due to the nature of the job (e.g., a four-color print job cannot be completed on a two-color machine). Further, certain presses are poorly suited for a particular job (e.g., a short production run on a machine with high set-up costs). Determining which of the thousands of customer orders to allocate to which printing press each year is a critical factor in managing costs in an increasingly competitive marketplace. Standard Register traditionally established its order-allocation rules centrally; however, it often made *ad hoc* tactical decisions based on local press capacity, geographic proximity to a customer, and marketing incentives.

Standard Register's goal was to develop order-allocation strategies that would result in cost reduction or improved customer service; however, ideally both cost and service could be improved simultaneously. Standard Register's operating environment is very complex with many possible product feature combinations. Consequently, in order to create a manageable project scope, the study summarized here focused on Standard Register's high-volume lithographic print business that accounted for annual revenues of $350 million and whose productive capacity was geographically dispersed around the United States at six plants. Each of these plants produced two types of orders: make-to-order and make-to-stock (shown in Figure 8-2). In the make-to-stock environment, orders (for steady customers via managed accounts) are produced based on forecasts and shipped to warehouses in anticipation of demand. In the make-to-order environment, customers place orders with sales associates. The orders are then entered into the order-entry system and routed based on the sales associate's best estimate of which facility in his/her geographic proximity will best be able to meet the promised due date.

In the production process, printing presses represent one of the most, if not the most, significant equipment investments at Standard Register. Pre-print and post-print operations are important to the successful delivery of a quality print job; however, the costs associated with printing operations make this a particularly important phase of the production process. As a consequence, it is of particular importance to allocate jobs to the printing press that will minimize the cost to the firm's production-distribution network. The research team involved in this project wanted to demonstrate to Standard Register's top management team the value in updating its operations strategy to move toward a paradigm of supply chain optimization. Specifically, the research team sought to identify a segment of Standard

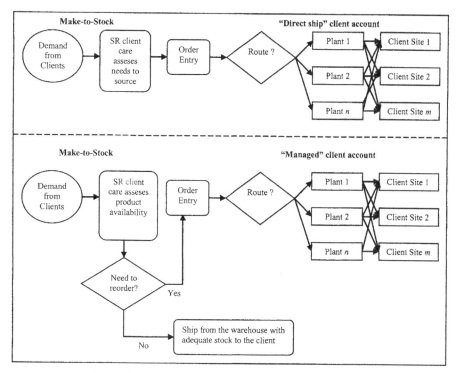

Figure 8-2   Production-distribution network based on order type

Register's business that was representative, high value, and separable from other business segments such that the results of that segment alone were meaningful and able to be implemented as a part of a greater improvement effort that would address the broader challenges facing the firm. The focus of the present study then was to help the firm to rethink the logic it employs in allocating incoming orders in this specific segment of its lithographic print business to its production-distribution network.

Three classic operations research techniques were selected to help Standard Register update its production-orders allocation logic to achieve its goals of reducing total landed costs and improving customer service. The research team elected to model system operation based on quarterly analyses. Historical data suggested the presence of a demand pattern that was stable between quarters and which followed a regular pattern within quarters. This allowed the research team to provide a meaningful analysis to management without needlessly increasing model complexity or data requirements.

Further, the team put into place a policy constraint that required Standard Register to keep as much work in-house as possible (subcontracting was a last-resort source of capacity) since the company was known to have excess capacity on its printing presses. First, regression analysis was used to predict the cost and efficiency attributes of each printing press based on print jobs of different types. Three important parameters needed for the subsequent optimization and simulation models included estimates of production time, production cost, and transportation cost. Once this statistical analysis was completed, an optimization model was developed based on the classic assignment integer programming problem. The purpose of this optimization model, implemented in Lindo Systems Incorporated's What'sBest ®, was twofold. On the one hand, it was important to demonstrate to top management that allocating a job to the press closest to a customer or to the press that is cheapest will not always lead to a minimization of total costs for the system, or lead to satisfactory customer service levels. It is often necessary to be more flexible in determining which press is the best one to satisfy a particular customer order.

The primary purpose of the optimization study, however, was to identify a set of strategic rules for allocating jobs to an appropriate printing press that could later be tested in the simulation model and in practice. The optimization model identified the optimal assignment of a set of jobs in a deterministic environment with an aggregated level of press capacity. Considering aggregate capacity alone in assigning a set of jobs to the firm's printing presses fails to recognize the dynamic planning environment found in practice. It is one thing to have enough aggregate capacity to handle demand for a particular planning horizon (e.g., a quarter); however, it is quite another thing to have capacity available when it is needed. For example, if two orders arrive that require a particular printing press at the same time, then a situation exists where, at least for some period of time, there is insufficient capacity to meet demand.

The third management science tool used in the study, a simulation model, was created using ProcessModel ® to capture the dynamic order routing, production, and shipment through order-fulfillment process. The simulation model allowed for the system dynamics associated with stochastically occurring customer-order arrivals, press availability and downtimes, and the constraint of meeting variable expected ship dates given press outages and availability. To create a baseline for comparison, Standard Register's current operating practices were modeled to benchmark actual system performance in terms of historical costs and customer service levels. In total, six business rules for assigning incoming customer jobs to printing presses were then explored:

1. *Historical only*—make assignments following historical assignments
2. *Least cost only*—assign the incoming job to the least cost press only
3. *Least cost only, then subcontract*—assign the incoming job to the least cost press only, but if it is not available subcontract the job
4. *Lowest cost available, then subcontract*—assign the incoming job to the lowest cost press available, but if no presses are available subcontract the job
5. *Optimization recommendation only*—assign the incoming job to the optimal press recommendation only
6. *Optimization recommendation first*—then lowest cost available, then subcontract

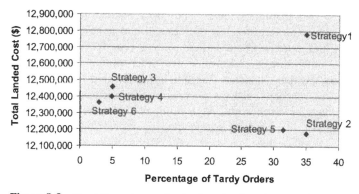

Figure 8-3   Simulation outcomes for the six order allocation strategies for one quarter

The simulation was run, using each of the preceding rules as a scenario, for two quarters. The first quarter of operation was used as an initialization period in order to build up queues at the presses, and performance data were collected based on the second quarter of operation. The simulation was designed to track, in addition to cost, such variables as order-cycle time, percentage of orders late, average lateness of orders, and the number of orders assigned to the first choice press. Figure 8-3 graphically shows the simulation results in terms of total landed cost and the percentage of customer orders tardy for a one-quarter simulation using the six order allocation strategies. Order tardiness is a measure of customer service. It indicates the percentage of customer orders that were delivered after the promised delivery date.

The results of the optimization study indicated an opportunity to reduce total landed cost of the high-volume lithographic print business by approximately 3.5% compared to its current job allocation strategy, representing a potential savings of approximately $1.5 million. The simulation confirmed the results of the optimization study indicating a better order allocation strategy would lead to a reduction in total landed costs. The cost savings identified in the simulation study were approximately 70% of the optimization study's predicted cost reduction. Neither the optimization-recommended job allocation strategy (Strategy 5) nor the least-cost press strategy (Strategy 2) provided sufficient flexibility to improve both cost and customer service (based on percentage of tardy orders). However, Strategies 3, 4, and 6 resulted in significant predicted cost savings accompanied by a significant reduction in the percentage of tardy customer orders.

The business impact of the study described above is impressive. The combination of scientific analyses, their accuracy and power, and an effective managerial briefing led the Executive Leadership Team at Standard Register to rethink its operations strategy. One of the most significant changes resulting from the study is that job assignments and subcontracting decisions are no longer made within marketing. Job allocations are now truly based on centralized knowledge of capabilities, costs, and press availability across the production network. Standard Register has also established a new senior level executive position, Chief Supply Chain Officer, which reports directly to the CEO and is, among other things, responsible for order-routing decisions for the entire organization. Further, Standard Register has been able to consolidate and redistribute its press capacity and, as a consequence, increased its aggregate capacity while lowering fixed costs. As a result of these changes, at least in part, Standard Register increased its net income by more than $31 million from 2004 to 2005 on sales that increased only $10 million over the same time period.

Visteon Visteon is a major supplier of integrated automotive systems to Ford Motor Company. In addition to producing newly manufactured goods, it also remanufactures product modules for sale in the automotive aftermarket.[2] Visteon uses the term *remanufacturing* (reman) for transforming used units (cores) into refurbished units that satisfy precisely the same quality standards as newly manufactured units. This process is highly labor intensive and highly variable, with processing times dependent upon the age and wear present in the input core. Visteon's Lamosa, Mexico, plant remanufactures such product families as rack-and-pinion steering gears, recirculating ball-nut gears, and power steering pumps for several models of Ford cars, trucks, and recreational vehicles. This plant suffered the effects of heavy demand and supply fluctuations, resulting in periods of severe capacity shortages. Management at Visteon was interested in analyzing options to deal with this capacity problem. Two basic options being advocated by process engineers at the plant were: 1) reconfigure the plant and 2) duplicate the existing reman line. Management was reluctant to make a significant capital investment and was uncertain what represented the best option for reconfiguring the reman line. The focus of the study being summarized here was the rack-and-pinion reman line.

The rack-and-pinion reman line was structured like a repair facility. Cores were first stripped of worn-out (non-reuseable) parts, then cleaned, inspected visually, tested for straightness and cracks, machined, assembled, tested for functionality, and finally packed and shipped. The line's order of operations can be characterized as shown in Figure 8-4. There was little product variation in the rack-and-pinion reman line. The top eight product families accounted for more than 80% of the demand. Variation in core quality was problematic and was responsible for losses in through-put rates.

The research team, after visiting the Visteon facilities, and collecting and analyzing data identified four bottleneck operations: cleaning, polish/wash, the final assembly station (Assembly D), and functional testing. These are indicated in bold in Figure 8-4. Management already had improvement plans in place to alleviate the

---

[2] Adapted from Kekre S, Rao US, Swaminathan JM, Zhang J. "Reconfiguring a remanufacturing line at Visteon, Mexico." *Interfaces* 2003;33(6):30–43.

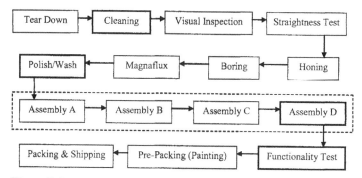

Figure 8-4   Rack and pinion reman line

bottleneck stations with the exception of Assembly D. The research team believed it could alleviate the bottleneck represented by Assembly D by reallocating tasks and workers across the four-station assembly line (represented by the dashed box in Figure 8-4).

This four-station assembly line was operated in an asynchronous mode. This means that when a part was completed at one station it was moved to the next station as soon as that workstation was available. The reman line operated with no storage buffers between workstations due to the existence of a zero-inventory policy. Workers were assigned to a specific workstation (e.g., Assembly A) and trained to operate the fixtures and equipment required to complete their assigned tasks. Workers who inspected an incoming core tested for a variety of defects, and this inspection determined the processing-time requirements at the four downstream assembly stations. Processing times at each station varied between 500 and 750 seconds. Task times at these stations could also be positively or negatively correlated.

Several factors determine the production capacity of a reman line: line and process design (layout, inventory buffers), line operation (line balancing, worker training), and characteristics of the operating environment (processing time variability and correlation, core condition). Visteon could exert little control over processing time variability since it could not control the quality of the incoming cores. Given these factors, Visteon wanted the research team to address the following two questions: 1) Should Visteon use a number of short lines (e.g., a single workstation) or a few long lines (e.g., three workstations)? and 2) What improvements could Visteon make by dynamically assigning tasks to workstations?

The dynamics of the reman line the research team was studying differ significantly from those of a traditional manufacturing line. These differences exist along two primary dimensions. In a traditional manufacturing line, usage quantities of the parts required to produce a product are known. For example, if you are building a bicycle, you know that you will need two wheels, one handlebar, one seat, etc. In remanufacturing, these usage quantities are not known until a core is examined because usage will be dependent upon the condition of the core. A good quality core will require fewer parts than a more worn core. Second, in traditional manufacturing environments, processing times are known with a high degree of accuracy. Reman lines, in contrast, exhibit highly variable processing times in repair and finishing, and these times are not known until the core is tested.

In order to address the questions posed by Visteon management, the research team developed an optimization and simulation model. The team modeled Visteon's reman assembly operation as a serial production line in which remanufacturing the product form the core required M tasks. The serial line was segmented into N stations, with each station being assigned a subset of the M required tasks. In Visteon's current configuration, the rack-and-pinion line has four workstations. Before a core enters the assembly line, workers test the core and determine a list of tasks required for assembly along with estimated task processing times. Consequently, it was assumed that the processing times for tested cores were known with certainty.

The first challenge was to determine how to configure the assembly lines; that is, how many production stations to have in each assembly line and how many assembly lines to implement. Obviously, each configuration would affect tooling, worker-training, and worker-retention costs. Tooling costs were of minimal importance at Visteon since the assembly process was largely manual and located in Mexico. A backlog of profitable customers suggested that a modest capital investment for reconfiguration would have a short payback period. The most problematic issues were related to human resources. Visteon was willing to address the human resource issues only if the research team was able to demonstrate a minimum increase in throughput of 20%.

The second challenge was, based on line configuration, to determine which of the M tasks to assign to the N workstations in order to balance the workload. The research team explored two methods for balancing the workload: static line balancing (SLB) and dynamic line balancing (DLB). Under static line balancing tasks are assigned to different workstations based on expected task-processing times prior to the testing of a core. In dynamic line balancing, tasks are assigned to stations for each individual core after testing the core and determining the processing time of all tasks. This scheme could be implemented by, after the initial testing of a core, attaching a process-routing slip indicating the tasks to be executed and assigning required tasks to the different stations of the assembly system.

It is not a simple matter to evaluate the performance of asynchronous, serial systems with random processing times and finite inventory buffers. At this point, there is no known closed-form expression for the expected throughput of Visteon's four-station assembly line. Consequently, the research team conducted a computational study based on the data it had collected during its visit to the Visteon automotive parts remanufacturing facility. To conduct the computational experiments, a discrete-event simulator was developed in C programming language. For each line configuration, 12,000 jobs were generated. The first 2000 jobs were used to initialize the system. Throughput and standard error in the simulation were calculated based on the 2001st through 12,000th jobs.

Based on the results of the simulation study, the research team was able to develop recommendations to Visteon's management that were relatively simple to implement and which met their stated performance requirements. The team recommended that the reman line be reconfigured from a four-station assembly line to a single-station assembly line. This change provided an expected increase in throughput of between 25 and 30% given Visteon's current task-processing times. Additionally, a single-station assembly line does not require balancing. This simplifies the assignment of work. The only drawback to the proposed solution is the cost to cross-train workers. Cross-training can mitigate the undesirable effects of employee absenteeism and provides management with a more flexible work force. The single-station assembly line also makes a single worker team responsible for a specific product's assembly quality. Feedback from the Visteon remanufacturing plant indicated that management was pleased with the results of this simulation study. A significant improvement in reman throughput on the rack-and-pinion gear line was facilitated with a relatively minor investment. This initial study represented a proof-of-concept for the reconfiguration other reman lines at the Lamosa, Mexico, plant.

This brief introduction to real-world applications of simulation to business was intended to provide the reader with a cursory exposure to the kinds of simulation that are appropriate in this environment. It demonstrated that well-conceived simulation studies with concrete results can be used to drive change within organizations at all levels of planning and execution: from operational to tactical to strategic. Further, it should be noted that simulation is applied across the broad spectrum of organizations, from manufacturing to service.

It is reasonable to expect that increasing numbers of business professionals will be involved with, or commission, simulation studies to address business problems and opportunities. Simulation has become increasingly accessible to individuals at all levels of the business organization. Some relatively simple simulation models will be created and implemented to facilitate what-if analyses using plug-ins to off-the-shelf spreadsheet modeling programs. Other simulations will be more sophisticated but will rely upon special-purpose software packages developed in visual interactive modeling systems [2].

This increased visibility of simulation in business will prove to be both boon and bane. Problems previously untouched due to a, real or perceived, lack of appropriate decision support tools will be increasingly addressed. Businesses will enjoy the financial rewards associated with solving such problems. Unfortunately, managers' expectations for simulation results will inflate to unreasonable levels and simulation will be force-fit as the methodology of choice inappropriately. It will be the simulation professional's responsibility to manage these expectations and to make sure that simulation is applied judiciously such that businesses will continue to reap the rewards of well-conceived and executed simulation studies.

MEDICAL MODELING AND SIMULATION

C. Donald Combs

Medical simulation has now become an accepted methodology for educating future medical practitioners and for providing ongoing training and assessment for practicing professionals. The *Residency Review Committee for Surgery* recently voted to require simulation training in surgical residency programs and the *American College of Surgeons* has established an accreditation process for surgical simulation/education centers [23]. In addition, the *National Board of Medical Examiners* has included a mandatory simulation component in their licensing exam using standardized patients in an Objective Structured Clinical Examination [24, 25]. Further, it has been proposed that the six general competencies developed by the *Accreditation Council for Graduate Medical Education* should be expanded to include a seventh procedure-based competency that could utilize simulation training for assessment of initial procedural competency and ensure the maintenance of competency throughout practitioners' careers [26]. Why are these changes happening now, and what are the key imperatives in medical education, healthcare delivery, and medical ethics that are contributing to the acceptance of medical simulation?

The pedagogical approach utilized in many of the nation's medical schools has not changed markedly since the revisions resulting from the Flexner report a century ago. The limitations of this apprentice-style approach, often characterized as the *see one, do one, teach one* methodology, have become more apparent because of its expense and the economic realities of modern medical education. This pedagogical methodology requires a low faculty-to-student ratio that is becoming less economically feasible as medical school enrollments have increased in recent years without corresponding increases in medical education funding and because of limitations in the subsidies for education that used to come from research and patient care revenue [27–32]. Effective use of simulation promises to help faculty members to become more productive and to leverage their teaching time across a larger number of students. Medical simulation can also help to reduce the subjective element of medical education by incorporating objective metrics for assessing student and resident performance.

Undergraduate medical education has been a *hit or miss* proposition as it relates to the development of particular procedural skills in students. For example, a medical school might schedule two suturing training sessions a year, but attendance at these sessions is usually voluntary and there is no guarantee that upon graduation a student would have had any significant experience with suturing. In recent years, there has been an increasing push by accrediting organizations to hold schools accountable for documenting the specific procedural and cognitive skills that are taught in each clinical clerkship experience. This has resulted in the rapidly increasing use of simulation because there can be no guarantee that a particular patient on which to practice a particular skill will just happen to be available to each student during their six-week clerkship experience. Simulation can help to mediate the vagaries of which types of patients are available and thus provide a more standardized training experience that lends itself to the documentation requirements of accrediting bodies such as the *Liaison Committee on Medical Education* (LCME) [33].

In graduate medical education, such as in a surgical residency, there are an ever-increasing number of surgical procedures to be mastered by surgical residents in the finite timeframe of a residency; a timeframe that has been further constrained by the 80-hour work week limitation for residents. In the past it was not unusual for residents to work 90–100 hours per week! The reason residencies are currently so lengthy (three to seven years) is because the variability in individual training experiences is reduced the more time one spends in residency-training. For example, the likelihood that a particular surgical resident will be faced with a particular surgical situation x increases as the time t spent in a residency is increased. Residency accrediting bodies also want to ensure that each resident has had a specific portfolio of experiences during their training. Simulation affords the opportunity to provide a simulated clinical experience for the resident should the genuine experience not happen to present itself during the time that the particular resident is on-duty. The result is that simulation helps to assure a more uniform educational experience for the resident. Simulation technology also offers the opportunity to incorporate a *hands-on* component into recertification examinations that historically have been predominately paper and pencil examinations given to practitioners who completed their formal training years before. The challenge of recertification is to ensure not only an increase in knowledge, but also a continued ability to use current knowledge through demonstrated competence in practice. The use of simulation in the recertification process can provide added safety for the public and assurance that there has not been a degradation of the physician's skills since the time of training and initial certification [34–39].

If medical simulation is to realize its full potential in medical education, then the simulators will have to be grounded in the specific requirements of the various curricula. The Association of American Medical Colleges (AAMC) has outlined the general requirements for undergraduate medical education during the four years of medical school. Those requirements include the physician's ability to understand the scientific basis of medicine; their ability practice medicine; their knowledge of the structure and function of the body; their engagement in a life-long study to remain current in the practice of medicine; their ability to provide care for a patient; their understanding of the effectiveness of therapeutic options available for a patient; and their ability to communicate with a patient [40]. These statements serve as a specification of the requirements for medical simulators to be incorporated into the first professional training curricula undertaken by medical students.

Following medical school, medical students enter graduate medical education programs called residencies and fellowships. The *American Council on Graduate Medical Education* has likewise developed a specification of general competencies that residents in each graduate medical education program are required to demonstrate. Those requirements include six areas of expertise that define the specific knowledge, skills, and attitudes required and provide educational experiences: 1) patient care, 2) medical knowledge, 3) practice-based learning and improvement, 4) interpersonal and communication skills, 5) professionalism, and 6) systems-based practice [41].

Once graduate medical students complete their residency training programs and any subsequent subspecialty fellowship training, they enter professional practice and will continue to practice for 30–40 years. This period of active professional practice will be significantly affected by the discovery of new medical knowledge, advanced diagnostic and therapeutic procedures, and increasing accountability for the outcomes of the patient care services they provide. Additionally, there will be economic pressures to increase their personal productivity by seeing more patients in ever-briefer periods of time, especially since the average face-to-face time during a physician-patient encounter is now about seven minutes; 15 years ago it was 11 minutes [42]!

The need to gain and maintain competence and to demonstrate proficiency in the use of new technologies make a compelling case for the increased use of simulation technologies in medical education and medical practice. Dr. S.B. Issenberg has suggested a framework for thinking about how medical simulators might be usefully deployed in this regard. The emphasis on repetition, measurement of performance, and feedback promise to strengthen and to standardize important components of medical education. We will return to these features later when the current state of research on medical simulation is discussed [43].

A Brief History of Medical Simulations

The *Advanced Initiatives in Medical Simulation* (AIMS) 2004 report entitled, "Building a National Agenda for Simulation-based Medical Education," assessed the full range of publicly funded medical simulation projects and made recommendations about the future growth and development of the industry [44]. Funded by the U.S. Army's Telemedicine and Advanced Technology Research Center (TATRC), the report categorized current approaches to medical simulation and their potential benefits (see Table 8-10). This category provides a useful framework for considering the historical development of medical simulation.

Medical simulations are almost two millennia old. The first simulations of living humans were developed by dissecting dead humans. Galen was a prominent Greek physician who practiced dissection. His writings on anatomy influenced physicians from 200 A.D. to 1500 A.D. and his emphasis on blood-letting as a remedy for almost any ailment persisted well into the 1800s. Although Galen wrote mostly about human anatomy, his dissection and research used pigs and apes, particularly the Barbary ape, and was therefore often erroneous.

In about 1489 Leonardo da Vinci produced a series of anatomical drawings that were far better than Galen's or anything else previously attempted. Over the next 25 years he dissected about 30 human corpses, many of them at a mortuary in Rome, until in 1515 Pope Leo X ordered him to stop. His drawings included studies of bone structures, muscles, internal organs, the brain, and even the position of the fetus in the womb. His studies of the heart suggested that he was on the verge of discovering the concept of the circulation of the blood.

Shortly thereafter, a young medical student, known to history as Vesalius, attended anatomy lectures at the University of Paris in the mid-16th century. Lecturers at the university explained human anatomy, based on Galen's work of more than

Table 8-10   Simulation Tools and Approaches Used in Simulation-based Medical Education

Type	Description
Low-tech simulators	Cadaver, models or mannequins used to practice simple physical maneuvers or procedures
Simulated/ standardized patients	Actors trained to role-play patients for training and assessment of history taking, physicals, and communication skills
Screen-based computer simulators	Programs to train and assess clinical knowledge and decision making, e.g., perioperative critical incident management, problem-based learning, physical diagnosis in cardiology, acute cardiac life support
Complex task trainers	High-fidelity visual, audio, touch cues, and actual tools that are integrated with computers. Virtual reality devices and simulators that replicate a clinical setting, e.g., ultrasound, bronchoscopy, cardiology, laparoscopic surgery, arthroscopy, sigmoidoscopy, dentistry
Realistic patient simulators	Computer-driven, full-length mannequins. Simulated anatomy and physiology that allows handling of complex and high-risk clinical situations in life like settings, including team training and integration of multiple simulation devices

1000 years earlier, as assistants pointed to the equivalent details in a dissected corpse. Often the assistants could not find the human organ as described, but invariably the lecturers decided that the corpse, rather than Galen, was incorrect. Vesalius decided to dissect human corpses himself and report exactly what he found. His approach was highly controversial, but his skill led to his appointment in 1537 as professor of surgery and anatomy at the University of Padua. In 1540 Vesalius gave a public demonstration of the inaccuracies of Galen's anatomical theories, which were still the orthodoxy of the medical profession. As noted, Galen did many of his experiments on apes. Vesalius, however, compared the skeletons of human beings and apes. He was able to show that in many cases Galen's observations were correct for the apes, but not humans. Vesalius then set about correcting Galen's errors in a series of dissections and drawings. Vesalius published his great work entitled, *De humani Corporis Fabrica* (The Structure of the Human Body), in 1543. There are seven volumes including magnificent woodcut illustrations. The book was an immediate success and allowed readers to peer beneath their own skins, in strikingly clear images of what lies hidden.

In 18th-century Italy anatomical waxworks were developed to represent the human body. Life-sized, colored, three-dimensional, soft, and moist-looking, these waxworks offered compelling replicas of the living body. For those who wanted to see something of anatomy without attending a dissection, the development of these anatomical waxworks offered a realistic alternative. One of the first physicians to use wax models was the French surgeon Guillaume Desnoües (1650–1735); he specialized in educational models. In the early, 18th century Ercole Lelli (1702–1766) and his pupil Giovanni Manzolini (1700–1755) founded a school of anatomical wax modeling at the Institute of Sciences of the University of Bologna. A second school of anatomical modeling, founded at the Florentine Museum of Physics and Natural History, later known as *La Specola*, eventually contained over 1400 anatomical wax models including 19 life-sized male and female figures, and was the first museum of its kind open to the public. Wax models did not remove the need for dissection, however. Modeling them was a painstaking process and hundreds of corpses, which quickly decayed in the Italian heat, had to be dissected to provide the subject matter for just one model [45–47].

These wax models evolved into working models of the human body fashioned in leather and wood. For example, in the mid-1700s an obstetrical teaching device was fashioned to contain an extraordinary, doll-like fetus *in utero*. These models also simulated the rudiments of childbirth. One model came from the Hospital del Ceppo in Pistoia, near Florence, and others exist in Bologna and Paris. The *baby* could be placed in any position and delivery demonstrated [48].

These basic approaches to medical simulation, models and dissections, were dominant until the 1960s, when more sophisticated approaches such as highly trained actors and computer-based mannequins began to appear. Actors trained to mimic patients, called **standardized patients** (SP), have gone through many metamorphoses as the process itself has been refined since its inception in 1963. There have been many other names attempting to describe this phenomenon: actors programmed to portray a patient, patient instructor, patient educator, professional patient, surrogate patient, teaching associate, and the more generic term *simulated patient*. All of these terms refer to a person who has been carefully trained to take on the characteristics of a real patient in order to provide an opportunity for a student to learn or be evaluated on skills *firsthand*. While working with the standardized patient the student can experience and practice clinical medicine without jeopardizing the health or welfare of real patients. The value is derived from the *experience* of working with a patient. Using SPs takes learning a step beyond books and a reliance on paper and pencil tests. It puts the learning of medicine in the arena of clinical practice—not virtual reality, but veritable reality—as close to the truth of an authentic clinical encounter as one can get without actually being there, because it involves a living, responding human being.

The expression *standardized patient* was coined by the Canadian psychometrician Geoffrey Norman, who was looking for a designation that would capture one of the technique's strongest features: *the fact that the actor could play exactly the same role for many students*. The term was adopted and generally accepted in the 1980s, when the focus of medical education research using simulated patients began to address clinical performance evaluation. The SPs offer students an opportunity to come face-to-face with the totality of the patient, with their *stories*, physical symptoms, emotional responses to their illness, attitudes toward the medical profession, stresses in coping with life, work and family. In other words, almost everything a real patient brings to a clinician, overt and hidden, allowing students to go about the *process* of unfolding all that they need to know from the veritable interaction with the patient in order to assist that person to heal [49].

The limitation, of course, is that the standardized patient does not generally have the physiological and anatomical traits of the sick patient they are portraying. So, there comes a point when the learner is unable to see, hear, or touch real symptoms of disease and injury. This shortcoming led to a second, contemporary approach to medical simulation—**mannequins**. The earliest mannequins were plastic models that were slowly modified to take advantage of the improving capabilities of embedded computers.

While not computer driven and having relatively limited functionality, the *Resusci®-Anne mannequin*, a relatively simple plastic model used for practicing heart compressions and mouth-to-mouth resuscitation, is the progenitor of the current mannequin simulators. Created in the late 1950s, this mannequin was designed by Asmund Laerdal, a successful Norwegian manufacturer of plastic toys [50]. Most readers will be familiar with Resusci®-Anne from the basic CPR training offered by organizations such as the Red Cross and American Heart Association. In wide use by the early 1960s, Resusci®-Anne marked the first modern medical simulator to impact medical education.

SimOne was a starting point for true computer-controlled mannequin simulators, particularly for simulation of the entire patient. Conceived by engineer Dr. Stephen Abrahamson and physician Dr. Judson Denson at the University of Southern California in the mid-1960s, it was built in collaboration with Sierra Engineering and Aerojet General Corporation. Abrahamson, in a video acceptance of an award from the Society for Technology in Anesthesia, described the idea as originating from Aerojet's need to develop peacetime applications of its capabilities in the face of diminishing military funding before the escalation of the Vietnam conflict. The initial concept of replicating anesthesia machine functions quickly evolved to one of recreating more of the entire patient. After meeting with rejection from the National Institutes of Health and military funding sources, the project to build a prototype was supported by a three-year, $272,000 grant from the U.S. Office of Education. The simulator was a remarkably lifelike mannequin, controlled by a hybrid digital and analogue computer. It had many high-fidelity features: the chest was anatomically shaped and moved with breathing, the eyes blinked, the pupils dilated and constricted, and the jaw opened and closed. To a limited extent it was used for training and to conduct some primitive experiments about efficacy. SimOne did not achieve acceptance and only one was constructed because the then-current computer technology was too expensive for commercialization. But, equally important, the market for approaches to medical training other than an apprenticeship model was nonexistent.

Harvey is a full-sized mannequin developed in the late 1960s that remains in use and currently simulates 30 cardiac conditions. It is the earliest example of the modern concept of a part-task trainer for medical skills training. It was first demonstrated in 1968 at the American Heart Association Scientific Sessions by Dr. Michael Gordon of the University of Miami Medical School under the title of Cardiology Patient Simulator. The simulator displays various physical findings, including blood pressure by auscultation, bilateral jugular venous pulse wave forms and arterial pulses, precordial movements, and auscultatory events that are synchronized with the pulse and vary with respiration. Harvey is capable of simulating a spectrum of cardiac disease by varying blood pressure, breathing, pulses, normal heart sounds, and murmurs (to learn more see http://www.cpmemedmiami. edu).

Important contributors to the history of modern mannequins were the development of mathematical models of the physiology and pharmacology of drugs in anesthesia. This development served a dual function: allowing mannequins to evolve into screen-based simulators for different applications and providing the underlying concepts in modeling physiology that were needed to support hands-on simulators with automatic controls. Several computer-based simulations of various aspects of anesthesia have been developed. For example, J.H. Philip created a program for teaching uptake and distribution of anesthetic agents, which he called *GasMan*(r) and J. Sikorski developed a computer-based simulation for instructing anesthesia residents in managing intraoperative events.

More complete models of human physiology have enabled higher-fidelity, more realistic simulations. Building on research in physiological modeling, Dr. N.T. Smith and colleagues at the University of California at San Diego (UCSD) conceived a multicompartment model of human physiology and pharmacology that formed the basis for *SLEEPER*, a screen-based simulator. Intended primarily to teach physiology and pharmacology, this simulator consisted of a fairly complex system requiring more computing power than was then available in desktop computers. SLEEPER evolved into a broader application, *BODY*™, which was marketed first by Marquette Medical Systems and currently by Advanced Simulation Corporation (to learn more see http://www.advsim.com/biomedical/body-simulation.htm).

At the Stanford University Medical School–affiliated Veterans Affairs Palo Alto Health Care System, Dr. David Gaba and colleagues fabricated what is probably the first prototype of a mannequin simulator developed for investigating human performance in anesthesia. The original version, used for experiments and training in early 1987, was numbered *CASE 1.2* (Comprehensive Anesthesia Simulation Environment). It combined commercially available waveform generators and virtual instruments. For example, a noninvasive blood pressure monitor on a Macintosh Plus computer with a commercially available mannequin created a *patient* whose vital signs could be manipulated to simulate critical events. Placed in a real operating room, this was the beginning of the high realism, physical simulation environment. The next version, *CASE 2.0*, contained a model of cardiovascular physiology running on a Transputer (a microprocessor chip designed for parallel processing with other such chips), making it partially model driven (as opposed to *CASE 1.2*) and also an early example of parallel processing.

At around the same time that *CASE 1.2* was being created, a multidisciplinary team at the University of Florida, Gainesville, led by Dr. Michael Good and mentored by Dr. J.S. Gravenstein, developed the Gainesville Anesthesia Simulator (GAS). It arose from an interest in training anesthesia residents in basic clinical skills. The project began with the capability for diagnosis of faults in anesthesia

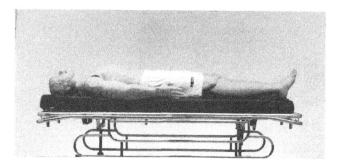

Figure 8-5   Human patient simulator

machines, in which controllable mechanical failure modes were embedded. Attaching a lung simulator to the machine extended the training challenge. This was developed into a complete mannequin, aimed primarily toward diagnosis of single source critical events in anesthesia. The mannequin had a sophisticated lung model that mimicked uptake and distribution of anesthetic gases. Later versions incorporated a system for automatically recognizing drugs as they were injected. In contrast to the more instructor-driven manual operation of *CASE 1.2*, this software enabled sequences of physiological changes both predefined and in response to actions of the trainer and trainee.

GAS was licensed to Loral Data Systems Inc., which later spun off the simulator product to a new company, Medical Education Technologies, Inc., (METI). The METI product (see Figure 8-5) was dubbed the Human Patient Simulator™ (HPS). A pediatric version, Pediasim™, and a simpler and more portable model, the ECS™, were introduced in the late 1990s and early 2000s, respectively.

These new-generation mannequins are complex and they afford a range of training capabilities for both individual and group training to address an increasing array of clinical situations. Four such mannequin simulators are described below.

NOELLE, the birthing simulator from Gaumard, is a full-sized articulating female mannequin with an articulating birthing baby and a full-sized intubatable and cyanotic newborn (see Figure 8-6). NOELLE allows the students to care for the mother and fetus before delivery and the mother and neonate after delivery. Features include intubatable airway with chest rise, intravenous (IV) arm for medication, practice leopold maneuvers, multiple fetal heart sounds, automatic birthing system, multiple head descent and cervical dilation, multiple placenta locations, replaceable dilating cervix, articulating fetus with placenta, and four delivery speeds and ten fetal heart rates (to learn more see http://www.gaumard.com).

The *Trauma®Man System* is widely used in military, emergency medical seervice (EMS), and other trauma educational courses for procedures that include cricothyroidotomy, chest tube insertion, pericardiocentesis, diagnostic peritoneal lavage, and IV cutdown (to learn more see www.simulab.com/TraumaSurgery.htm).

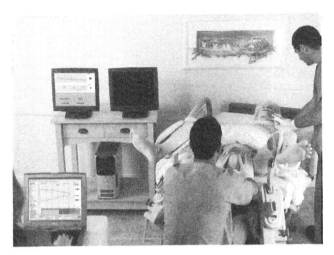

Figure 8-6 Noelle birthing simulator

CentraLine Man is an anatomically correct human patient simulation with landmarks that allow the user to practice a wide range of subclavian, supraclavicular, and internal jugular techniques. The simulator differentiates the arterial and venous blood to show positive or negative response. There are self-sealing veins and skin for multiple practices and a hand pump that provides a pulse to the arterial vessels (to learn more see www.simulab.com/CentraLineMan.htm).

A third class of medical simulators involves trainers oriented to simple or complex tasks such as suturing, endoscopy, and laparoscopic surgery. A listing of some of these tasks or procedural simulators may be found in Table 8-11 and a brief description of two such simulators follows.

The *LapVR™ Surgical Simulator* (see Figure 8-7) by Immersion Medical was designed by reviewing the Fundamentals of Laparoscopic Surgery (FLS) curricula endorsed by the Society of American Gastrointestinal Endoscopic Surgeons (SAGES). The system's hardware and software interface simulates laparoscopic surgery and provides virtual-reality training in the essential skills required of laparoscopic procedures. The LapVR™ surgical simulator comes packaged as a complete system with selected software module(s) and height-adjustable cart, monitor swing, foot pedal, camera, and tools (to learn more see www.immersion.com/medical/products/laparoscopy/index.php).

The Suture Tutor (see Figure 8-8) is designed as a resource to help convey the essential skills required for making skin incisions and suturing in a variety of methods. The topics covered in the simulation are safe handling of materials, instrument management, planning and performing a linear incision, interrupted suturing, subcuticular continuous suturing, and knot tying. Each compact and portable trainee kit contains the CD-ROM with single user software license and all the relevant instruments, surgical consumables and synthetic soft tissue models necessary for

Table 8-11 Simulators by Earliest Identified Date of Reference

Simulator	Date of first publication
Fibre-endoscopy	1987
ERCP	1988
Colonoscopy	1990
Endoscopic trainer	1993
Laparoscopic surgical simulator	1994
Hysteroscopy	1994
Hollow organ closure	1994
Total hip replacement	1995
Ophthalmic simulator of laser photocoagulation	1995
Ophthalmic surgery simulator	1995
Intravenous catheter insertion	1996
Otolaryngology	1996
Laparoscopic surgery	1997
AAA endovascular repair	1998
Virtual simulator for inferior vena cava filter placement	1998
Sigmoidoscopy	1998
Shoulder arthroscopy	1999
Surgical suturing	1999
Breast biopsy simulation	1999
Transurethral prostatic resection	1999
PC-based interventional cardiology simulator	2000
Bronchoscopy	2001
Upper gastrointestinal endoscopy	2003

hands-on skills training and skills reinforcement (to learn more see www. limbsandthings.com/uk/products.php?id:239).

Ways of Classifying Medical Simulations

A simple classification of medical simulations divides simulators into those based on physical models, those that use computers to create illusions of reality, and those that combine the two models [51].

Advances in materials technology have led to dramatic improvements in the realism of physical simulation through models and many human tissues can now be recreated with a fair degree of accuracy. A range of relatively inexpensive models is now available and most hospitals and medical schools have a skills center where learners from a range of disciplines can practice regularly. Procedures commonly taught in this way include urinary catheterization, venepuncture, intravenous infusion, and wound closure. Such physical models are also useful for practicing relatively simple surgical procedures such as the removal of cysts and lipomas.

Figure 8-7 Lap VR simulator

Figure 8-8 The Suture Trainer

Computer simulations are an attractive alternative, especially as improvements in processing power make them increasingly realistic. Key principles govern simulation in fields ranging from surgery to entertainment. As a general guide, the more complex a simulated procedure is and the more variables it offers, the greater the computing power it requires. There is, therefore, a trade-off between high visual fidelity on the one hand and the ability to interact with the program on the other.

Virtual reality (VR) has emerged as a powerful technique within healthcare. Originating in the space program in the early 1990s, VR is now becoming established in many branches of medicine. Described as a collection of technologies that allow people to interact efficiently with three-dimensional computerized databases in real time using their natural senses and skills, VR combines a convincing representation of an organ system or body region with the means to work with that image as if it really existed. Such interaction can include manipulating, slicing, cutting, and *flying through* systems of the body sometimes using complex *haptic devices* that allow the user to get a real-time sense of touch.

Hybrid simulators combine physical models with computers often using a realistic interface (such as real diagnostic or surgical instruments) to bridge the gap between a mannequin and a computer. This avoids some of the technical difficulties associated with reproducing the feel of instruments and of human tissue while still allowing access to the advantages of computer simulation. Such hybrid environments are likely to become increasingly important as complex technologies such as surgical robots move into mainstream practice. Computer-based **procedural simulations,** whether on their own or within hybrids, have great potential as learning tools. The following section will consider such simulators in more detail, using a simple taxonomy to summarize the current state of the art (see Table 8-12).

Precision placement simulators allow the learner to practice the skills of accurate manipulation and precise positioning. The task is usually to insert an instrument or needle along a straight line, as in venepuncture, spinal anesthesia, or lumbar puncture. The visual background is almost static and can therefore be made extremely realistic, even with modest computers. Such simulators are already well developed

Table 8-12 A Simple Taxonomy of Simulators (adapted from Satava 2007) [52]

Skill	Manual requirement	Examples
Precision placement	Direct needle/instrument to a point	Intravenous needle insertion, lumbar puncture
Simple manipulation	Guide a catheter, endoscope or ultrasound probe	Angioplasty, colonoscopy, bronchoscopy, abdominal ultrasound
Complex manipulation	Perform a single complex task	Bowel or vascular anastomosis, MIST-VR, LapSim
Integrated procedure	Perform multiple tasks of entire procedure	Anesthesia simulation, laparoscopy procedures

and are entering the mainstream of educational resources, especially at undergraduate level. Examples include the *CathSim* by Immersion and *Suture Tutor* by Limbs and Things.

Simple manipulation simulators are currently among the most convincing available. They allow the learner to practice manipulating an instrument in response to a visual display. Procedures, which include endoscopy and diagnostic ultrasound, use an authentic interface (an actual fiberoptic endoscope or ultrasound probe) in conjunction with a computer-generated display that mimics the view obtained during a clinical procedure. A range of pathologies can be presented, creating varying degrees of difficulty for the learner.

A range of simple manipulation simulators is now commercially available and literature in support of their effectiveness is growing. Lower gastrointestinal endoscopy (flexible sigmoidoscopy and colonoscopy) is now well established, and upper gastrointestinal endoscopy modules are being expanded to include procedures such as endoscopic retrograde cholangiopancreatography (ERCP). Training in flexible bronchoscopy has recently been validated and similar simulators for urological investigation are currently being developed.

Complex manipulation simulators are designed for practicing more complex surgery. The realism and range of these simulations is limited by their extremely high demands for computer power. Despite dramatic improvements in technology, the goal of fully realistic complex manipulation simulators remains distant. However, it is now becoming possible to practice a range of complex procedures (such as anastomosis and the management of limb trauma) as well as isolated components of technique. Many of these simulators have been developed for procedures relating to minimally invasive surgery. These procedures require the surgeon to manipulate instruments whose tips cannot be seen directly, relying on a computer display to gain a picture of the patient's anatomy.

Early complex manipulation simulators required learners to grasp and transfer objects on the screen, using simple laparoscopic instruments. The *Minimally Invasive Surgical Trainer* system (MIST) used three-dimensional geometrical shapes without attempting surgical realism.

The current generation of manipulation simulators allows learners to practice a range of generic laparoscopic skills such as instrument navigation, tissue grasping, simple dissection, clipping of blood vessels, and intracorporeal knotting. These simulators are much more convincing than their geometrical precursors and more accurately represent bleeding and deformation of tissue. Time-limited scenarios are designed to reproduce some of the pressures of real-life operating such as unpredicted hemorrhage. Detailed performance data is recorded and can be presented in a variety of formats for feedback and assessment.

Research—Current and Future

Several recent studies point to an increasing body of evidence supporting medical simulations as effective training tools. One study, *Virtual Reality Training Improves*

Operating Room Performance: Results of a Randomized, Double-Blinded Study, was conducted by researchers from Yale University and Queen's University Belfast. It states that surgeons who train on the surgical Procedicus MIST simulator performed 29% faster, made six times fewer errors, and were five times less likely to injure nontarget tissue, gallbladder or liver, when performing surgical removal of a gallbladder (called laparoscopic cholecystectomy) in human patients [53, 54].

Another study, *Transatlantic Medical Simulation,* described an effort to use mannequins for remote training. Advanced training using *Human Patient Simulators* (HPS) is, for the large part, unavailable for the majority of healthcare providers in rural, remote, and less-developed regions of the world, either due to their separation from the major medical education centers or significant fiscal austerity. Remote access to HPS based on the Application Software Provider principles may provide the solution to this problem. The medical ASP (augmented standardized patient) concept proposed and developed by MedSMART has been subjected to an extensive qualitative and quantitative international test conducted among France, Italy, and the United States. Two SimMan HPSs (made by Laerdal) were used with one unit based in Michigan and one in France. While the French site had both remote and hands-on access to the simulator, the Italian site could access the HPS only remotely. Simulator visualization was provided by four remotely operated cameras at each HPS site. HPS-generated vital signs were transmitted to each site together with the interactive simulator control panel using a communications hub at the MedSMART facility in Ann Arbor. At the end of the training program, a Likert scale-based assessment test was given. The trainees showed statistically significant ($p < 0.03$–0.05) improvement in all testing measures. The Likert scale questionnaire revealed overwhelming satisfaction with the simulation-based distance training even when the access to the simulator was only remote (Italy). Confidence was also significantly improved.

This study validated for the first time the concept of simulation-based, fully interactive transatlantic medicine. The experiment proved that training based on advanced technologies transcends barriers of distance, time, and national medical guidelines. Hence, international simulation-based distance training may ultimately provide the most realistic platform for a large-scale training of emergency medical personnel in less developed countries and in rural/remote regions of the globe [55].

Gaba and colleagues have taken a very important step into the future. Their research indicates that simulation can be used to assess individual physician preparedness to manage an extremely low-frequency, life-threatening event encountered in practice. The fundamental problem with such events is that their rarity almost guarantees an uneven training experience across large numbers of residents. The result is that many residents enter practice without ever having witnessed any number of infrequent but not rare diseases or clinical conditions [56].

For more than a decade, as shown in a fourth study, advancing computer technologies have allowed incorporation of virtual reality (VR) into surgical training.

This has become especially important in training for laparoscopic procedures, which often are complex and leave little room for error. With the advent of robotic surgery and the development and prevalence of a commercial surgical system (da Vinci™ robot; Intuitive Surgical®, Sunnyvale, CA), a valid VR-assisted robotic surgery simulator could minimize the steep learning curve associated with many of these complex procedures and thus enable better outcomes. To date, such a simulation product does not exist; however, several agencies and corporations are involved in making this dream a reality (to learn more see http://www.mimic.ws/MIMIC-dV-Trainer-Flyer-Ltr-Feb08.pdf) [57].

Issenberg *et al.*, conducted the first *Best Evidence Medical Education* (BEME) systematic review of the research on the features and uses of medical simulations that were judged to lead the effective learning. The authors assessed the research based on the appropriateness of the study design, the implementation of the study, the appropriateness of the data analysis, and the quality of the conclusions and recommendations. Based on their review of over 100 articles, they reached two important conclusions: 1) the research literature on the use of medical simulation is generally weak, and 2) despite the weakness there is sufficient evidence to identify the features of simulations that are most likely to lead to effective learning.

The quality of the published research on medical simulation is generally weak. That said, Issenberg's review of the research literature found that the weight of the best available evidence suggests that high-fidelity medical simulations do, in fact, facilitate learning under the right conditions (see Table 8-13). These findings support the concept that the best simulators:

- *Provide feedback*—51 (47%) journal articles reported that educational feedback is the most important feature of simulation-based medical education
- *Involve repetitive practice*—43 (39%) journal articles identified repetitive practice as a key feature involving the use of high-fidelity simulations in medical education
- *Integrate with the curricula*—27 (25%) journal articles cited integration of simulation-based exercises into the standard medical school or post-graduate educational curriculum as an essential feature of their effective use
- *Possess a range of difficulty levels*—15 (14%) journal articles address the importance of the range of task difficulty level as an important variable in simulation-based medical education
- *Involve multiple learning strategies*—11 (10%) journal articles identified the adaptability of high-fidelity simulations to multiple learning strategies as an important factor in their educational effectiveness
- *Capture clinical variation*—11 (10%) journal articles cited simulators that capture a wide variety of clinical conditions as more useful than those with a narrow range

Table 8-13 Features of Simulations that Lead to Effective Learning

Variable	Feature	Description
Teaching Resources/ strategies	Feedback	Feedback provided during the learning experience is the most important feature of simulation-based education to promote effective learning
	Repetitive practice	Learners should engage in focused, repetitive practice where the intent is skill improvement, not just idle repetition
	Range of difficulty level	Learners should engage in skills practice across a range of difficulty levels, beginning with basics and advancing to progressively higher difficulty levels based on objective measurements
	Multiple learning strategies	Simulation-based learning strategies should include but not be limited to: instructor-centered formats, small group tutorials and independent study, depending on the learning objectives being addressed
	Clinical variation	Simulations should represent a wide variety of patient problems to provide more sampling than simulations that only cover a narrow patient range
	Controlled environment	Simulations work best when embedded in controlled educational settings where (unlike real clinical environments) learners can make, detect, and correct patient case errors without negative consequences
	Individualized learning	Educational experiences should be reproducible and standardized for individualized learner (or team) needs where learners (or teams) are active participants, not passive bystanders
	Defined outcomes/ benchmarks	Educational goals should have tangible, objective measures that document learner progress in terms of training benchmarks
	Simulator validity/realism	The simulation and the behavior it provokes come close to, but never exactly duplicate, clinical challenges that happen in genuine patient care contexts
Curricular integration	Curricular integration	Simulation-based education experiences are a routine feature of the normal educational schedule and are grounded in learner performance evaluation

- *Occur in a controlled environment*—10 (9%) journal articles emphasized the importance of using high-fidelity simulations in a controlled environment where learners can make, detect, and correct errors without adverse consequences

- *Utilize individualized learning*—10 (9%) journal articles highlighted the importance of having reproducible, standardized educational experiences where learners are active participants, not passive bystanders

- *Define expected outcomes*—7 (6%) journal articles cited the importance of having clearly stated goals with tangible outcome measures that will more likely lead to learners mastering skills;
- *Possess validity*—4 (3%) journal articles provided evidence for the direct correlation of simulation validity with effective learning [58, 59].

While simulation research needs improvement in terms of rigor and quality, high-fidelity medical simulations have already demonstrated their potential to be educationally effective and to complement the clinical education gained in patient care settings. As Kneebone noted, there are four key advantages of simulator-based learning. First, the training agenda can be determined by the needs of the learner, not the patient. Learners can focus on whole procedures or specific components, practicing these as often as necessary. Second, because the environment is safe, learners have *permission to fail* and to learn from such failure in a way that would be unthinkable in a clinical setting. This gives an opportunity to explore the limits of each technique rather than having to remain within the zone of clinical safety. Third, simulators can provide objective evidence of performance, using their built-in tracking functions to map a learner's trajectory in detail. An increasing range of metrics are being developed and validated, offering potential for formative and summative assessment. Fourth, the capacity of simulators to provide immediate feedback in digital form offers a potential for collaborative as well as individual learning [60].

Using simulation in a medical education setting is not without its challenges, however. Some studies are beginning to explore barriers to the use of simulation and are finding that the most common barrier for staff is the *lack of free time*. The most common barrier for trainees is the *lack of training opportunity* because of the clinical workload. Compared with trainees, staff clearly tend to perceive more barriers and identified lack of free time and financial issues as significant barriers.

Other perceived barriers included the stressful and intimidating environment of high-tech medical care, the fear of an educator's or peer's judgment, and the fear of an inaccurate reflection of one's own clinical ability. It is recognized that a simulated scenario may be stressful, may trigger strong emotions, and may increase the number of errors committed by the participant. The issues of having one's performance analyzed and critically reflected on can be very daunting. Nonetheless, this barrier serves to highlight the strength of high-fidelity simulations that can provide a safe venue for learners to commit errors and to reflect on them [61].

When mannequin-based simulation work began in the mid- to late 1960s, there was a strong belief among those early researchers that simulation would become important to healthcare over the following decade. Now in the fifth decade of computer-based simulation work, most researchers realize that research and development of medical simulations is still in the early phases of what will surely be a several decades-long process of embedding simulation into the fabric of healthcare. A number of observers have asked: *What is the next big invention in simulation?* With a broader view they might better ask: *What is the next big step for simulation?*

Although there is an abundant need for technology development to make simulations of greater veracity and for more applications, the biggest step for simulation going forward will not be technological, but organizational. That is, even with today's technologies there is an enormous amount that can be accomplished with simulation that is not being done because the institutional mechanisms for providing it are immature [62].

In a lead editorial in *Simulation in Healthcare*, Glavin summarized the results of a 2006 conference focused on the potential of medical simulation. Involving 65 participants with broad expertise in medicine and in simulation, the conference focused on answering four questions:

- How can medical simulation contribute to the education of trainees?
- What is the role of simulation in evaluating a trainee's general competency?
- How should we develop a research agenda to evaluate simulation?
- How should simulation technologies be developed and managed within and across institutions?

Four days of intensive discussion on these questions led to four recommendations that serve as a useful framework for thinking about the continued evolution of medical simulations:

- Medical simulation should be thoughtfully used and supported as a complement to current teaching methods for medical students, resident trainees, and faculty.
- The integrated use of various modes of simulation should be a priority in simulation efforts.
- Medical simulation research should pursue performance-based and patient-centered outcomes, using robust quantitative and qualitative measures.
- Advocates of medical simulation should give a high priority to multi-center collaboration as a means to stimulate the use of simulation technology and to study its efficacy [63].

Some specific areas for further research and development of medical simulations range from the level of cellular functioning to the integration of many anatomical and physiological components in a manner to simulate the workings of the entire human body—the virtual human. As complex a task as this integration will be, some of the major components that will be included in the virtual human simulators of the future are:

- Predictive modeling of specific biological systems for applications such as medical therapeutics and biomedical research. For example, predictive models of generic cell types such as red blood cells, eukaryocytes, and prokaryocytes could be used to screen the effects of novel drugs in pharmacological research. Similarly, patient-specific models could be used to predict the effect of novel pathogens on the respiratory systems of individual patients.

- Synthesis of diagnostic imaging, modeling, and simulation with therapeutics in real time to improve healthcare delivery. For example, matching of a patient's diagnostic images during surgery with a probabilistic atlas of normal anatomy encompassing the range of human variation could enhance accuracy and improve outcome.

- Validated and accurate simulations of major organs, organ systems, and interrelated organs integrating anatomy, physiology, biomechanical properties, cell biology, and biochemistry. These should include integrated models of the vertical organization of some of the major organs (heart, lung, muscle) as well as horizontally integrated models of major physiological systems (circulatory, respiratory, immune). Visualization and simulation technology should be available to move seamlessly between different spatial resolutions (molecular to organ level) and different temporal states (development through aging, varying physiologic state) within an integrated simulation.

- Creation of *body-double*, patient-specific image models that will serve as a repository for diagnostic, pathologic, and other medical information about a patient. These will serve as a three-dimensional (3D) template for enhancing communication between patient and physician, and provide a reference framework to examine pathologic and age-related changes that occur over time.

- High-fidelity medical simulation for training and accreditation. These simulators will support true user interaction with simulated human organs, including validated physical and physiological properties, such as real-time tissue deformability, realistic bleeding, and accurate haptics. High-bandwidth access will facilitate distributed visualization and simulation of models for medical education and research and development applications [64].

The practice of healthcare involves a variety of individuals with differing levels of education and training. From graduation to licensure to certification and recertification, the focus of the legal and professional regulatory system is on the individual. Proof of competence has historically been based on 1) the successful completion of a course of study ranging from some months to as many as 11 years of post-collegiate study, 2) the successful completion of national and/or state written exams, and 3) some process of continuing education that may or may not involve a periodic re-examination by either the state or the profession or both. This process of demonstrating competence holds generally true for laboratory technicians, physical therapists, pharmacists, dentists, nurses, physicians, administrators, and the manifold public and allied health workers who together constitute the U.S. health workforce. Recent innovations in the process have often focused on ways of assessing continuing competence in practice in addition to assessing the knowledge base. These innovations have arisen from concerns about the capacity of practitioners of all types to be knowledgeable about the plethora of emerging diagnostic and therapeutic technologies, and about their ability to continue to practice effectively during decades-long careers. The preceding narrative has clearly indicated the growing role that simulation is playing in the training of individual practitioners. As important as that is, however, the future of medical simulation will necessarily evolve from the mere training of individuals to the training of individuals in teams and in shared decision-making with the complex social and technological context of practice.

Effective processes for assuring the continuing competence of individuals is, of course, necessary for maintaining the quality of the U.S. healthcare system. As many observers have noted, however, individual competence is not sufficient to ensure either patient safety or high-quality patient care. The ever-increasing specialization of knowledge, the practical limits on how many hours a week can be safely worked, and the fact that patient health problems require competencies held by practitioners of different professions combine to require additional competence in working as a member of a team. Communication, cooperation, and the recognition and integration of disparate skill sets along with patience, perseverance, and shared responsibility and accountability are central attributes of effective team members. Many reform proposals over the years have urged medical and health education programs to do a better job of incorporating these concepts into their curricula. Similar calls for professional certification, licensure, and recertification programs to incorporate an emphasis on multi-disciplinary teamwork have not yet been heeded widely or systematically [65].

Unfortunately, as the striking similarity of the reform proposals demonstrates, only relatively modest changes in the central processes of medical and health professions education have occurred during the past century. Until individual competence is supplemented by team competence, both in education and in certification, the types of quality and safety problems that have been well documented in the U.S. healthcare system are inevitable. That many organizations, hospitals, group practices, and insurers are beginning to force change by requiring some evidence of team competence as a condition of the right to work in particular settings and to be reimbursed is again evidence of how the professions, the educators, and the regulators lag real-world practice. This then is the major emerging challenge of medical simulation: *developing realistic contextual training simulations that require both individual competence and group competence in the complex, urgent situations that constitute the healthcare setting.*

As many scholars of expertise have noted, the most important identifiable factor separating the elite performer from others is the amount of deliberate practice. This includes practice undertaken over a long period of time to attain excellence as well as the amount of ongoing effort required to maintain it. Deliberate practice has been defined as the opportunity to tackle a well-defined task with an appropriate difficulty level for the particular individual, informative feedback, and opportunities for repetition and corrections of errors [65]. In the future, the deliberate practice available through medical simulation must evolve to include decision-making within the context of teams of differently trained practitioners working with urgency.

Finally, of course, is the need for a systematic body of research to focus on the validity and training transfer effects of medical simulation. The aspiration of medical education and practice to be based on science and the rigorous scientific research on the efficacy of medical simulation as an alternative to current practice in education and training is just beginning to emerge. Far more simulation research to date has been focused on the analysis and modeling that is necessary to develop simulations. In the future, however, the increasing investment in and use of medical simulations will necessarily lead to more questions about whether the simulations do what their designers say they do, and even if so, do they make the practice of medicine safer and more likely to result in a beneficial outcome for patients? These are the important questions that have yet to be answered definitively.

SOCIAL SCIENCE MODELING AND SIMULATION

John A. Sokolowski and Catherine M. Banks

As we have learned, models are representations of structures or systems. Simulations are models with specific inputs entered by the modeler. The outputs are what are observed as the simulation runs. In the social sciences the inputs to a model are the attributes needed to make the model match a social environment. The outputs to this simulation are the behaviors of the model [66]. Social science modeling is important because it allows for the substitution of human capabilities (such as in a virtual operating room) and it assists in discovery as the researcher proffers a hypothesis then analyzes the meaning of the outcome. As you recall, M&S allows us to control the actual object of interest because we are using the model, not the person or occurrence. Simulation is a good way to understand social processes because it can be used to **predict** the features of an environment, such as the changing demographics of a region over the course of time; or **forecast** changes such as business trends. Social scientists typically conduct two types of broad investigation after determining a theory: **inductive reasoning**, which allows the theory to be tested via generalized observations and **deductive reasoning**, which tests the assumption of the hypothesis. Simulation presents the social scientist with a third type of investigation, one that starts with assumptions, then experiments to generate data, then allows for analysis of the data [66].

The social sciences make a significant contribution to the multidisciplinary application of M&S. Thus, it is important for you to note how social science modeling augments, supports, and provides various types of analyses to the mathematical formulation that goes with the M&S we have been discussing throughout this textbook. This chapter subsection will introduce you to the social sciences and the types of modeling typically done in those disciplines. To provide a real-world example of social science M&S, we have included a case study on modeling population movement during an insurgency. This case study will help you to appreciate the transitioning of qualitative analysis, typically done by social scientists, into the quantitative data needed to produce a system dynamics model that can represent population transfer.

What Are the Social Sciences and What Types of Modeling Do Social Scientists Do?

The **social sciences** are academic disciplines that study human aspects of the world—civilization—emphasizing the use of the scientific method in the study of humanity that includes quantitative and qualitative methods. There are many disciplines within the social sciences: anthropology, comparative theology (religious studies), criminal justice (and/or criminal science), economics, education, history, philosophy, political science (government), social psychology, social work, and sociology. These disciplines study human society and individual relationships *in* that society as well as individual relationships *to* that society. The quantitative and qualitative method of inquiry lends itself to different modeling methods.

M&S applications in the social science disciplines are becoming more and more common. It is apparent that not all modeling is quantitative or numeric, in fact, qualitative analysis introduces *ideas* into the model. Engaging M&S into social science analysis allows us to better understand the *what happened* and to explore the *what if*. Recall from Chapter 1 the history of M&S and its strong ties to the U.S. military. Today, those ties are even closer because the military has recognized its need to incorporate social science analysis into its training and decision-making simulations. The Department of Defense is now requiring the incorporation of geopolitics, culture, religion, and political economy variables to better understand how **Diplomatic, Intelligence, Military, and Economic (DIME)** factors affect real-time decision making. The U.S. military employs simulation to train all levels of its service personnel and to analyze complex policy and warfighting plans. Most of these simulations have represented only the military (force on force) aspects of warfare and they have ignored the political, social, and economic aspects that are vital to understanding *Effects-based Warfare* currently employed by combatant commanders.[3] As a result, a growing area of research for the U.S. military is centered on understanding the **Politics, Military, Economics, Social, Information**, and **Infrastructure (PMESII)** aspects of a region before finalizing its operational and tactical planning.[4] Without accurate models of these social science areas combatant commanders are unable to test effects-based strategies and plans beyond tabletop exercises. The social sciences are now integral to the military and organizations and institutions that need to understand the complexities of human behavior because simulations that contain social science models will much better represent real-world consequences.

There are many methods of social science modeling. Some of the most common are *statistical modeling*—a traditional method for the discovery and interpretation

[3] The military defines *effects-based warfare* as the application of armed conflict to achieve desired strategic outcomes through the effects of military force.

[4] Operational planning refers to warfare conducted at a headquarters level while tactical planning refers to that conducted on or near the battlefield or area of operation by those immediately engaged in the conflict.

of patterns in large numbers of events; *formal modeling*—a method that provides a rigorous analytic specification of the choices actors can make and how those choices interact to produce outcomes; and *agent-based simulation modeling*—a method allowing for the observation of aggregate behaviors that emerge from the interactions of large numbers of autonomous actors. Chapter 1 introduced you to *multi-agent modeling*—models that focus on the way in which social behavior emerges from the actions of agents; *behavior modeling*— a model of human activity in which individual or group behaviors are derived from the psychological or social aspects of humans; and *social network modeling*—a model of social behavior that takes into account relationships derived from statistical analysis of relational data. Let's discuss in greater detail some of these social science modeling methods.

Agent-Based Modeling

Agent-based models are dynamic models that imitate the actions and interactions among the *units of analysis*.[5] **Agent-based modeling** focuses on these units of analysis, or agents, and the sequence of actions and interactions of the agents over a period of time. Agents may represent people, organizations, countries—any type of social actor. These actors may act in parallel, may be heterogeneous, and may learn from their actions. Each agent responds to the prior action of one or more of the other agents or the environment in the model (or system). This, in turn, produces an extended and often emergent sequence of behaviors that can be analyzed for different things. The action of the agent can be regarded as a variable; inaction of the agent is also a variable and it can be considered hostile, neutral, or information seeking.

Agent-based modeling is intrinsically social in that the actions and characteristics of the agents are influenced by the actions and characteristics of the other agents in the social system. Specific to agent-based modeling are the computer-based simulation programs in which the agents are active. This type of modeling is generally bottom-up because the simulation begins by focusing on actions of the individual agent. (Opposite to this is the system dynamics model that takes a top-down approach and focuses on the entire system then narrows down to the agent level.)

To conduct agent-based modeling one begins by defining the basic behavior of an agent. This behavior may be captured in many ways. The most often used method is through a series of simple rules that the agent must follow. These rules help describe the fundamental goals that the agent is trying to achieve. Probably the best-known example of an agent-based model is Craig Reynolds' *Boids* [67].[6] Boids simulates the flocking behavior of birds with three simple rules. We show those rules in Table 8-14.

With the application of these three simple rules (goals), Reynolds showed that flocking behavior of birds would emerge. This idea of emergence is central to agent-

---

[5] For our purposes we define *agents* as autonomous software entities that receive input from their environment and act on that input to achieve some goal or perform some action.

[6] Craig Reynolds *Boids* page can be found at http://www.red3d.com/cwr/boids/.

Table 8-14   Boids Behavior Rules

Rule	Explanation
Separation	Birds steer to avoid crowding local flock mates
Alignment	Birds steer towards the average heading of local flock mates
Cohesion	Birds steer towards the average position of local flock mates

based models. Because of the agent's autonomous, goal-seeking behavior, complex behaviors emerge that are not preprogrammed. This methodology is an effective way to simulate complex social behaviors through the application of relatively simple rules that each agent follows. The reader is invited to explore some of the implementations of agent-based models in the social science area found on the World Wide Web.[7]

Game Theory (Behavior Modeling)

Game theory modeling is tied closely to the problem of rational decision making. Decision makers, be they politicians, group leaders, military commanders, or chief executive officers, rarely find the ideal or rational solution to contend with problems they face. Game theory serves as a tool to study the interactions of individuals, also called *players*, in various contexts such as social dilemmas or combative situations. The intent of game theory is to observe the interactions between and among the players so that the decision makers can determine what they deem to be the best course of action.

There are two types of game theory: *cooperative game theory*, whereby the players can communicate to form winning coalitions, and *non-cooperative game theory*, which focuses more on the individual and their handicap of not knowing what the other players will do. Hence, this individual has to make a choice among the available options that are not won via coalition building [68]. As you can see, game theory facilitates the ability to analyze strategic behavior where there are conflicts of interest. Often, the result of this type of analysis allows the social scientist to categorize interpersonal behavior within a spectrum of cooperative or competitive behavior.

Many social science studies of cooperative and competitive behavior focus on the familiar notion of *Prisoner's Dilemma*—two players, Prisoner A and Prisoner B, are separated and each one is told things that are supposedly said by the other actor. These things may be true or untrue as it forces each player to make decisions or take actions without the certainty of what has actually been stated by the other. The dilemma is the uncertainty in which each prisoner finds himself. Should Prisoner A

---

[7] Multi-player games are possible but beyond the scope of this discussion.

cooperate or *defect*? To cooperate is to act in the collective interest of the two prisoners and not believe anything told him or her. Prisoner A may choose to defect if he or she believes Prisoner B has acted against him or her. Prisoner A's defection will serve to save his or her own skin. Defection is part of rational decision-making—*save yourself*; it is not part of the idealism of collective interest. Yet, human behavior is flexible as both cooperation and competition are equally engaged at different times depending on what conditions in which the actors finds themselves. Game theory serves to allow the social scientist to explore various policies to see which provides the best overall result. Optimizing ones payoff in light of an opponent's strategy is the goal.

Game theory was conceived by John von Neumann and Oskar Morgenstern as a method to study economic policies [69]. Others quickly realized its application to many areas, including politics. The Rand Corporation employed game theory to devise Cold War policies concerning nuclear weapons strategies.

To model a social conflict such as the Prisoner's Dilemma, the scientist must first devise a problem that revolves around a conflict between two players.[8] This scientist must then determine a set of payoffs or rewards that each player would receive depending on whether they cooperate or try to challenge one another. Each player's goal is to maximize their payoff given the expected action of an opponent. From the actions of each player the scientist can study the overall outcome of the game for each player and infer an optimum strategy or policy given the setup of the game.

The optimum strategy or policy is not always obvious. Even with a game as simple as the Prisoner's Dilemma there are many possible policies that can be played out over time. A famous challenge by Robert Axelrod invited scientists from all over the world to propose algorithms that would evolve an optimum policy [70]. Many such algorithms were tried and one known as *Tit for Tat* proved to the best overall strategy.

Social Network Modeling

Social network modeling helps us to understand the connections among people whether they be political leaders, specific groups, and/or cliques in organizations. It also allows for an explanation of a flow of information, or the spread of contagion, or the identification of outliers or individuals who are isolated from the group. Grouping patterns (algorithms) allow for the separating of large networks into smaller subsets. This draws closer the members who share identifying marks or attributes. Perhaps they are all from the same religious sect, the same age, or they have the same political ideology. Integral to social network modeling is the analysis or disciplined inquiry into the patterns of relationships that develop and exist among the members in the social system. Also included are the relationships among members at different levels of the analysis, i.e., person to person, groups

---

[8] *Game Theory: Analysis of Conflict* by Roger B. Myerson provides a more complete treatment of this subject and is provided as a reference for the student to further explore this area.

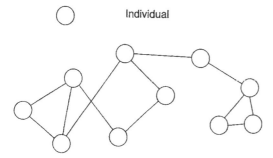

Figure 8-9 Social network model

to groups, etc. Because this type of modeling relies on actors that are concrete and observable, the relationships within the social network are usually social or cultural. These types of relationships bind together the actors or entities making them interdependent entities.

Social networks are characterized by nodes that represent the entities sharing some relationship, and links which connect the nodes together in this shared relationship. Figure 8-9 is a representative network.

The links may be one to one, many to one, or one to many. However, each node may only have one link associated with it or a node may originate or receive multiple links depending on how it relates to the other nodes in the network.

One example of social network use has been to examine the spread of disease during an epidemic. In this case each node represents 1) a person susceptible to the disease, 2) a person who has contracted the disease, or 3) a person who is immune from the disease. The links represent who has contact with whom and thus the possibility of transmitting the disease to a susceptible person. The probability of the susceptible person contracting the disease is determined by many factors that are represented as part of the network node. In this manner epidemiologists may simulate how the disease is spread. They may be able to investigate ways to limit its spread by using a social network model to experiment with different public policies to see which might be the most effective. Without a social network model, this experimentation would not be feasible.

System Dynamics

The founder of the system dynamics approach to modeling is Professor Jay Forrester of the Massachusetts Institute of Technology Sloan School of Management. Dr. Forrester, a professor of electrical engineering (University of Nebraska), conducted research experimental studies of organizational policy. He used computer simulations to analyze social systems and predict the implications of different models. This method came to be called system dynamics. **System dynamics** deals with the simulation of interactions between objects in dynamic (active) systems. He describes his approach in the following manner [71]:

"System dynamics combines the theory, methods, and philosophy needed to analyze the behavior of systems not only in management, but also in environmental change, politics, economic behavior, medicine, engineering, and other fields. System dynamics provides a common foundation that can be applied wherever we want to understand and influence how things change through time."

Forrester J.W. System Dynamics and the Lessons of 35 Years.
1991 Apr 29. Report nr D-4224-4.

This description makes clear why many social scientists look to system dynamics modeling to create a macro-level representation of a system that can address the interdependence of the actors, events, or variables within the system.[9]

To model a system using the system dynamics approach, one first develops a **causal loop diagram**. This type of diagram describes how various system variables relate to one another from a cause and effect standpoint. Figure 8-10 depicts a typical causal loop diagram.

This diagram shows how birth rate and death rate influence the overall level of a population. It consists of variables (birth rate, death rate, population, fractional birth rate, average life time) connected by arrows that describe the cause and effect relationship. The positive (+) or negative (−) sign at each arrowhead indicates how the dependent variable changes based on a change in the independent variable. For example, if the birth rate increases, there will be a positive effect on population. The large positive sign and negative sign inside their own loops indicate the overall type of loop with positive (+) representing a reinforcing loop and negative (−) representing a balancing loop.

One drawback of causal loop diagrams is that they do not allow for the accumulation of variable totals. This limitation is overcome by using these diagrams to develop stock and flow models. *Stocks* are accumulation points within the system that allow one to measure the amount of a variable at any given time. *Flows* are inputs and outputs of stocks that represent the rate of change of the stock. Figure 8-11 shows a typical stock and flow diagram derived from the causal loop diagram of Figure 8-10. Here, population is the variable we are interested in and whose total number is being tracked.

The following case study will serve to illustrate an application of system dynamics modeling to a real world problem. The purpose of the study was to better under-

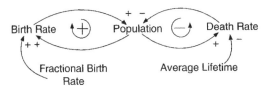

Figure 8-10   Influence diagram

---

[9] Here we define a *system* to mean a group of interacting or interdependent entities that form an integrated whole.

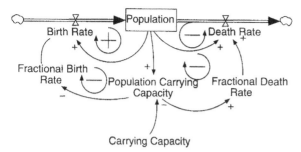

Figure 8-11   Stock and flow diagram

stand the behavior of a civilian population during insurgent and counter-insurgent activities via a review of tangible and intangible variables that affect human behavior and group dynamics.

Case Study: Modeling Population Movement During an Insurgency

Insurgencies are a part of man's quest for self-determination and self-governance. Throughout history we have observed repeated patterns of human behavior and that is most likely due to the fact that certain values do not change. In the ancient world the root cause of insurgencies, a strong desire for autonomy and self-governance, remains the same cause in the 21st century. This study sought to measure the behavior of a population during a period of insurgency by engaging system dynamics and mathematical modeling methods to characterize the structure of an insurgency and the causes or factors that influence the civilian population to join or oppose the insurgency.

To characterize insurgencies using a system dynamics approach requires an analysis of social behavior. Why do some individuals change their opinion of the insurgency and transition from unaffected (A) to affected (B) to engaged (C) in the insurgency? Those *affected* can be called dissidents, or nonviolent members of a population. To understand how members of the population shifted from one group to another we had to do a qualitative study of insurgencies. This allowed us to list factors that influenced the shift from A to B to C. There is an abundance of literature citing cases of insurgent activities as well as counter-insurgencies. States have varied in their approaches to counterinsurgencies. Comparing the intent of various insurgencies with the methods states have used to counter that activity allowed us to characterize, and then model insurgencies. Here are some of the variables we noted in the case study that affected population movement from A to B to C:

- history of the state or region
- religious divisions

- ethnic divisions
- relationship between the existing government and the civilian population
- relationship between the existing insurgents and the civilian population
- nature of the insurgency
- nature of the counterinsurgency
- loss of leadership within the insurgency
- perception of power on the part of the antagonists, the insurgents
- perception of power on the part of the protagonists, the counterinsurgents
- economic state of affairs
- civilian desire for security
- influence of neighboring countries

This list of factors was then binned into five general variables that influenced the movement of the population from one sub population to another.

- violent acts
- propaganda
- social network influence (describes the one-to-one relationships insurgents have with various members of the population and the insurgent's ability to influence those members to join the insurgency)
- loss of leadership
- government policies (policies act as catalysts to either placate or incite the population away from or towards insurgent behavior)

In counterinsurgent measures the government in power can set policy that goes straight to the heart of the civilian population because it touches four key areas: culture, religion, economy, and security.

With this initial qualitative assessment of insurgencies complete, the next step was to determine how to construct the system dynamics model. We had already completed the first step in creating this model by defining the three segments of the population, A-B-C, and determining the factors that influenced their movement. We viewed insurgency from a global perspective noting that insurgents as groups were not constrained by geographical boundaries or specific political systems. (This assumption allows for looking at insurgencies that arise from religious policies that have no tie to a specific government. The obvious example is the Al-Qaida insurgency driven by that group's perception of unacceptable policies against Muslims.) However, we did assume that an insurgency was targeted at a specific government or nation state as the focus of its effort.

As you can see from Figure 8-11 we must identify variables that influence the rate of change of our selected population. To identify these variables we must first decide on the populations to include in the model. At a minimum our model must include a background population representing all noninsurgents and a separate population containing the insurgents. We chose to characterize a population in three groups, the baseline susceptible population (S), dissidents (D), and insurgents (I). We assume that the susceptible population exists at a certain point in time with no influx of people being added. Dissidents may also exist and are characterized by nonviolent opposition to the policies of the government. Insurgents make up the remaining portion of the population and are characterized by those people with violent tendencies toward the government. Figure 8-12 represents the stock and flow model of our insurgency.

You will notice that it is possible to appease the dissidents to a point that they return to the general population. However, once a person becomes an insurgent he or she either stays in that group or is removed through attrition or some other means and does not return.

The discussion above identified four propensity factors that influenced the flow of insurgents: culture, religion, economy, and security. These insurgency propensity factors are manifested in a population through government policies in these four areas. Policies may have a positive effect on a population to refuse support of an insurgency. Or, the policies may have a negative effect, which will result in support of the insurgency. This influence is shown in Figure 8-13.

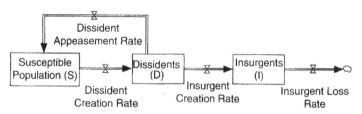

Figure 8-12   Insurgency model stock and flow diagram

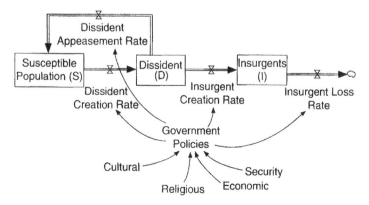

Figure 8-13   Insurgency model showing government policy influence

On the insurgent side we found four factors that contributed to controlling the flow rates in the above model. *Violent Acts* can be a rallying point for insurgents. They may persuade dissidents to join the insurgency or they may cause fear and tend to isolate the insurgent cause. *Propaganda* refers to information reaching the population and the outside world through printed, electronic, or video news means. Media use is a very significant factor in influencing a population's feelings and allegiance to the above groups. A *Social Network Influence* refers to the personal contact between individual and groups of insurgents and other members of the general or dissident population. Personal persuasion or coercion to join the insurgency is another contributor to the overall flow rate in the above system.

The only other factor that we broke out as a specific influence on the system was *Loss of Leadership*. In some cases where there are strong leaders and a tightly coupled insurgency organization, losing a leader figure may have a significant effect on the functioning of that effort. In loosely coupled insurgent cells, loss of leadership may be less significant. Figure 8-14 depicts our complete insurgency model.

As with any model, we could have added many subfactors under each of these main ones. However, we wanted to insure this model was broad enough to characterize all types of insurgencies. You can add more levels if you want to model specific insurgencies or include more detailed factors under some of the major factors.

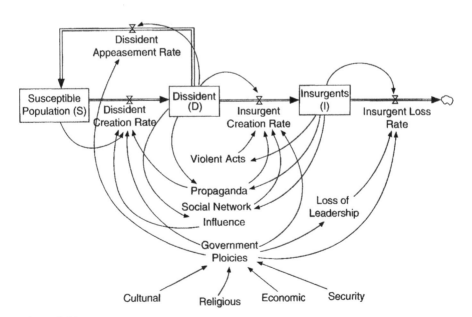

Figure 8-14   Full insurgency model

CONCLUSION

Let's review the four applications areas that have been introduced in this chapter. *Transportation M&S* is used to explore ways to meet the challenges of the escalating transportation woes. Transportation touches every day life of each member of society. Those who participate in providing these services do so with an emphasis on speed, safety, and reliability. The transportation system is multimodal because it has a variety of ways to move goods and people. The transportation community is employing innovative ways to manage congestion with current and future transportation challenges. Transportation M&S is the ideal tool to examine and proffer solutions to challenges that are part of Transportation Planning and Transportation Operations.

The discussion on *Business M&S* introduced us to the use of simulation in industrial organizations as a decision support methodology. Today, Business M&S is used in both manufacturing and service sectors playing a prominent role in a number of organizations as a result of the significant achievements of simulation. In Business applications, M&S allows the model developer to test the effects of change to system components or human behavior on system performance. This subsection focused on applications for the purposes of education and training in functional areas such as production, service, management, marketing, and finance.

Our next discussion focused on *Medical* applications. Medical M&S has a long history stemming from the ancient Greek practice of blood-letting to the anatomical diagrams of daVinci to waxworks to the present day use of standardized patients, haptic devices, and virtual reality. Medical simulation is now part of the teaching methodology for medical practitioners because it affords students the opportunity to train with a hands-on approach in the absence of a live patient. As we have learned the medical profession places very stringent requirements on its students such as those outlined by the Association of American Medical College and the American Council on Graduate Medical Education. Medical practitioners need to learn their discipline of study while gaining competency with the numerous new technologies available to them. The large body of medical literature stresses the importance of combining subject matter expertise and technical skill. While it is understood that nothing substitutes for practice in the study of medicine—it is now becoming widely accepted that medical simulation can provide an expanded variety of options for students to obtain that practice.

The last section of the chapter focused on *Social Science* applications. There are many disciplines in the social sciences and they all center on the study of human society and the individual relationships in that society and to that society. The four types of social science modeling methods commonly engaged are Agent-based Modeling, Game Theory, Social Network Modeling, and System Dynamics. Much of the research in the social sciences is qualitative, non-numeric, introducing ideas and data that are often hard to quantify. M&S is a tool that facilitates the translation of qualitative findings to quantitative values so that models can be developed to better understand human society.

KEY TERMS

multimodal	car following theory	forecast
travel demand model	system analysis	inductive reasoning
four-step model	acquisition and system	deductive reasoning
traffic analysis zone	acceptance	social sciences
impedance function	Monte Carlo simulation	DIME
Frank-Wolfe algorithm	Top Management	PMESII
TRANSIMS	Decision Simulation	agent-based modeling
trip chain	ITEC	game theory
capacity analysis	standardized patient	social network
capacity	mannequin	modeling
level of service	virtual reality	system dynamics
microscopic traffic	procedural simulations	causal loop diagram
simulation	predict	

FURTHER READING

COMBS CD. "Our nation today and in 2020: Are we preparing the health professionals we will need?" In *Factors Affecting the Health Workforce.* Holmes DE, Pryce Jones KE, eds., Washington, DC: Association of Academic Health; 2005.

GABA DM, RAEMER D. "The tide is turning: Organizational structures to embed simulation in the fabric of healthcare." *Simulation in Healthcare: The Journal of the Society for Simulation in Healthcare.* 2007;2:1–4.

GLAVIN RJ. "Simulation: An agenda for the 21st century." *Simulation in Healthcare: The Journal of the Society for Simulation in Healthcare.* 2007;2(2):83–85.

Institute of Medicine. *Crossing the Quality Chasm: A New Health System for the 21st Century.* Washington, DC: National Academy Press; 2001.

Institute of Medicine. *To Err Is Human: Building a Safer Health System.* Washington, DC: National Academy Press; 1999.

ISSENBERG SB. "The scope of simulation-based healthcare education." *Simulation in Healthcare: The Journal of the Society for Simulation in Healthcare.* 2006;1:203–208.

MYERSON RB. *Game Theory: Analysis of Conflict.* Cambridge, MA: Harvard University Press; 1991.

SIMON H. *The Sciences of the Artificial.* Cambridge, MA: The MIT Press; 1996.

SOKOLOWSKI JA, BANKS CM. "From Empirical Data to Mathematical Model: Using Population Dynamics to Characterize Insurgencies." In Proceedings of the 2007 Summer Simulation Multiconference. July 15–18, 2007. San Diego, CA, pp. 1120–1127.

ZIV A, et al. "Simulation-based medical education: An ethical imperative." *Academic Medicine.* 2003;78(8):783–788.

REFERENCES

1. SCHRANK D, LOMAX T. The 2007 Urban Mobility Report. Available from: Texas Transportation Institute, Texas A&M University, College Station, TX.

2. Bush aims at reducing air traffic congestion. November 2007. Available at: http://money.cnn.com/2007/11/15/news/economy/bc.apfn.airlinedelays.ap/. Accessed 15 March 2008.

3. BERNSTEIN M. The more trade grows, the worse U.S. port congestion becomes. *World Trade Magazine,* August 2006.

4. MTA–The Big Dig. Available at: http://www.masspike.com/bigdig/index.html. Accessed 15 March 2008.

5. ORUZAR J, WILLUMSEN L. *Modelling Transport*, 2nd Ed. New York: John Wiley and Sons; 1999.

6. RILETT LR. Transportation Planning and TRANSIMS Microsimulation Model: Preparing for the Transition. Transportation Research Record 1777; 2001. Available at: National Research Council, Transportation Research Board, Washington, DC.

7. Transportation Research Board. Highway Capacity Manual; 2000. Available at: National Research Council, Washington, DC.

8. MAY A. *Traffic Flow Fundamentals*. Upper Saddle River, NJ: Prentice-Hall; 1990.

9. DOWLING R, SKABARDONIS A, ALEXIADIS V. Traffic Analysis Toolbox Volume III: Guidelines for Applying Traffic Microsimulation Software; 2004. Available at: Federal Highway Administration, Washington, DC.

10. PARK B, WON J. Microscopic Simulation Model Calibration and Validation Handbook; 2006. Available at: Virginia Transportation Research Council, Charlottesville, VA.

11. ELDREDGE DL, WATSON HJ. An ongoing study of the practice of simulation in industry. *Simulation and Gaming* 2006;27(3):375–386.

12. ROBINSON, S. Discrete-event simulation: From the pioneers to the present, what next? *Journal of the Operational Research Society* 2005;56:619–629.

13. SALTZMAN RM, MEHROTRA V. A call center uses simulation to drive strategic change. *Interfaces* 2001;31(3):87–101.

14. KATSALIKAKI K, BRAILSFORD SC. Using simulation to improve the blood supply chain. *Journal of the Operational Research Society* 2007;58:219–227.

15. ANDREATTA G, BRUNETTA L, RIGHI L. Evaluating terminal management performances using SLAM: The case of Athens International Airport 2007. *Computers and Operations Research* 34:1532–1550.

16. NANCE RE, SARGENT RG. Perspectives on the evolution of simulation. *Operations Research* 2002;50(1):161–172.

17. FARIA AJ. Business simulation games: Current usage levels—An update. *Simulation & Gaming* 1998;29(3):295–308.

18. LEWIS MA, MAYLOR HR. Game playing and operations management education. *International Journal of Production Economics* 2007;105:134–149.

19. BERRY WL, MABERT VA. ITEC: An integrated manufacturing instructional exercise. *International Journal of Operations and Production Management* 1992;12(6):3–19.

20. BABICZ G. Simulation: Making it real. *Quality* 2003;42(4):90–91.

21. AHIRE SL, GORMAN MF, DWIGGINS D, MUDRY O. Operations research helps reshape operations strategy at Standard Register Company. *Interfaces* 2007;37(6):553–565.

22. KEKRE S, RAO US, SWAMINATHAN JM, ZHANG J. Reconfiguring a remanufacturing line at Visteon, Mexico. *Interfaces* 2003;33(6):30–43.

23. American College of Surgeons: Division of Education–Accredited Education Institutes. Accessed on January 3, 2008, from http://www.facs.org/education/accreditationprogram/list.html.

24. DILLON GJ, et al. Simulations in the United States Medical Licensing Examination™. *Qual Saf Health Care* 2004;13:i41–i45.

25. TAN GM, et al. Teaching first-year medical students physiology: Does the human patient simulator allow for more effective teaching? *Singapore Medical Journal* 2002;43(5):238–242.

26. American Council on Graduate Medical Education. General Competencies: Minimum Program Requirements Language. Accessed on February18, 2008, from http://www.acgme.org/outcome/comp/compmin.asp.

27. O'NEIL EH. The Third Report of the Pew Health Professions Commission. Critical Challenges: Revitalizing the Health Professions for the Twenty-First Century. Executive Summary 1995. Accessed on November 7, 2005, from http://www.futurehealth.ucsf.edu/summaries/challenges.html.

28. CHRISTAKIS NA. The similarity and frequency of proposals to reform U.S. medical education: Constant concerns. *Journal of the American Medical Association* 1995;274(9):706–711.

29. Institute of Medicine. *To Err Is Human: Building a Safer Health System.* Washington, DC: National Academy Press; 1999.

30. Institute of Medicine. *Crossing the Quality Chasm: A New Health System for the 21st Century.* Washington, DC: National Academy Press; 2001.

31. Institute of Medicine. *Health Professions Education: A Bridge to Quality.* Washington, DC: National Academy Press; 2003.

32. Institute of Medicine. *Priority Areas for National Action: Transforming Health Care Quality.* Washington, DC: National Academy Press; 2003.

33. Liaison Committee on Medical Education. Functions and Structure of a Medical School. Accessed on February 25, 2008, from http://www.lcme.org/functions2007jun.pdf.

34. SATAVA RM. The Future of Medical Simulation: Where We've Come, Where We're Going—Curricula, Criteria, Proficiency and the Paradigm Shift in Simulation. presented at *MODSIM WORLD 2007*, Virginia Beach; September 11, 2007.

35. CHAER RA, et al. Simulation improves resident performance in catheter-based intervention. *Annals of Surgery* 2006;244(3):343–352.

36. DHARIWAL AK, et al. Effectiveness of box trainers in laparoscopic training. *J Min Access Surg* 2006;3:57–63.

37. KORNDORFFER Jr. JR. Developing and testing competency levels for laparoscopic skills training. *Arch Surg* 2005;140:80–84.

38. REZNICK RK, MACRAE H. Teaching surgical skills—Changes in the wind. *The New England Journal of Medicine* 2006;355(25):2664–2669.

39. STURM L, et al. Surgical simulation for training: Skills transfer to the operating room. ASERNIP-S Report No. 61, 2007. Adelaide, South Australia.

40. Association of American Medical Colleges. Report 1: Learning Objectives for Medical Student Education-Guidelines for Medical Schools. Accessed February 22, 2008, from http://www.aamc.org/meded/msop/msop1.pdf.

41. American Council on Graduate Medical Education. General Competencies: Minimum Program Requirements Language. Accessed on February 18, 2008, from http://www.acgme.org/outcome/comp/compmin.asp.

42. COMBS CD. Our nation today and in 2020: Are we preparing the health professionals we will need? In Holmes DE, Pryce Jones KE, eds. *Factors Affecting the Health Workforce.* Washington, DC: Association of Academic Health; 2005.

43. ISSENBERG SB. The scope of simulation-based healthcare education. *Simulation in Healthcare: The Journal of the Society for Simulation in Healthcare* 2006;1(4):203–208.

44. Advanced initiatives In Medical Simulation. Building a National Agenda for Simulation-based Medical Education. Accessed on February 22, 2008, from http://www.medsim.org/articles/AIMS_2004_Report_Simulation-based_Medical_Training.pdf.

45. LUCIA D. Women, wax, and anatomy in the 'Century of Things'. *Renaissance Studies* 2006;21(4):522–529.

46. "Gray's Anatomy." Accessed on December 21, 2007, from http://en.wikipedia.org/wiki/Gray%27s_anatomy.

47. "History of wax anatomical models." Accessed on December 21, 2007, from http://medicina.unica.it/cere/mono01_en.htm.

48. LAWRENCE G. Tools of the trade: An obstetric phantom. *Lancet* 2001;358:1916.

49. WALLACE P. Following the threads of an innovation: The history of standardized patients in medical education. *Caduceus: A Humanities Journal for Medicine and the Health Sciences.* 1997;13(2):5–28.

50. COOPER JB, TAQUETI VR. A brief history of the development of mannequin simulators for clinical education and training. *Qual Saf Health Care* 2004;13(Suppl. 1):i11–i18.

51. KNEEBONE R. Simulation in surgical training: educational issues and practical implications. *Medical Education* 2003;37:267–277.

52. SATAVA RM. "Do or Do not, There is No Try," presented at 3rd Annual AIMS Conference, Washington DC, May 16, 2006.

53. VERSWEYVELD, L. Accessed on December 21, 2007, from http://www.hoise.com/vmw/02/articles/vmw/LV-VM-08-02-20.html.

54. GALLAGHER AG, CATES CU. Approval of virtual reality training for carotid stenting. *JAMA* 2004;292(24):3024–3026.

55. von LUBITZ DKJE, et al. Transatlantic medical education: preliminary data on distance-based high-fidelity human patient simulation training. *Stud Health Technol Inform* 2003;94:379–385.

56. GACA AM, et al. Enhancing pediatric safety: Using simulation to assess radiology resident preparedness for anaphylaxis from intravenous contrast media. *Radiology* 2007;245(1):236–244.

57. ALBANI JM, LEE DI. Virtual reality-assisted robotic surgery simulation. *Journal of Endourology* 2007;21(3):285–287.

58. ISSENBERG SB, et al. Simulation technology for health care professional skills training and assessment. *JAMA* 1999;282(9):861–866.

59. ISSENBERG SB, et al. Features and uses of high-fidelity medical simulations that lead to effective learning: A BEME systematic review. *Medical Teacher* 2005;27(1):10–28.

60. GOOD ML. Patient simulation for training basic and advanced clinical skills. *Medical Education* 2003;37(Suppl. 1):14–21.

61. SAVOLDELLI GL. Barriers to use of simulation-based education. Canadian *Journal of Anesthesia* 2005;52:944–950.

62. GABA DM, RAEMER D. The tide is turning: Organizational structures to embed simulation in the fabric of healthcare. *Simulation in Healthcare: The Journal of the Society for Simulation in Healthcare* 2007;2(1):1–3.

63. GLAVIN RJ. Simulation: An agenda for the 21st century. *Simulation in Healthcare: The Journal of the Society for Simulation in Healthcare* 2007;2(2):83–85.

64. HIGGINS G, et al. Final report of meeting Modeling & Simulation in Medicine: Towards an integrated framework. *Computer Aided Surgery* 2001;6(1):32–39.

65. COMBS CD, ALPINO RJ. The Emerging Imperative for Medical Simulation. Presented at MODSIM World 2007, Virginia Beach; September 12, 2007.

66. GILBERT N, TROITZSCH KL, eds. *Simulation for the Social Scientist.* New York: Open University Press; 2005.

67. REYNOLDS CW. Flocks, herds, and schools: A distributed behavior model. *Computer Graphics* 1987;21(4):25–34.

68. TROITZSCH KG, et al, eds. *Social Science Microsimulation.* Berlin: Springer; 1996.

69. von NEUMANN J, MORGENSTERN O. *Theory of Games and Economic Behavior.* Princeton: Princeton University Press; 1944.

70. AXELROD R. *The Evolution of Cooperation.* NewYork: Basic Books; 1984.

71. FORRESTER JW. System Dynamics and the Lessons of 35 Years. 1991 Apr 29. Report nr D-4224-4. Sloan School of Management, MIT.

Chapter 9

The Future of Simulation

R. Bowen Loftin

INTRODUCTION

Predicting the future has only one certainty—the more specific the prediction, the more likely it is to be wrong. This caveat applies to the contents of this chapter.

There are a number of past examples in which individuals (usually in panels) were asked to predict the "Future of Simulation." For many years, the annual Winter Simulation Conference has routinely offered a panel discussion on this topic.[1] In most cases the panelists dealt with "simulation in the small" not "simulation in the large." In this chapter we will deal primarily with "simulation in the large," more-over, the chapter will take the liberty of going beyond the normal boundaries of simulation and also deal with the human-simulator interface, some of the techno-logical underpinnings of simulation, and the relationship(s) of simulation to entertainment.

The future of simulation will, in this author's opinion, be determined, not by a systematic, well-coordinated effort of a body of academic researchers, rather it will be determined by forces beyond the control of any individual or small group of researchers—world events and public demand for entertainment will play the pre-dominant role in shaping the future of simulation. Nonetheless, there is a critical role for education in the future of simulation.

Finally, a disclaimer is in order: the ideas presented here, unless specifi-cally cited as those of another, are the author's own, and he will graciously accept the ridicule of the readers, both now and when he is proven wrong.

---

[1] See for example http://www.wintersim.org/prog03.htm#FS.

Principles of Modeling and Simulation: A Multidisciplinary Approach, Edited by John A. Sokolowski and Catherine M. Banks.

A BRIEF ... AND SELECTIVE ... HISTORY OF SIMULATION

By any measure, simulation has a long history. Humans are natural "simulationists." Young children will make or use models (dolls and other toys) to execute a simulation (play). Games such as chess (in the West) and Go (in the East) have served for hundreds of years as simulations of warfare. There is ample evidence in the historic record that live simulation has been employed for at least 2000 years.[2] Formal use of "wargaming" by the military became common in the 19th century. The beginning of the modern era of simulation, however, is usually associated with the advent of flight simulation in the early 20th century (Figure 9-1). Computer-based simulation began in the 1950s and, of course, is now commonplace. In almost every case simulation has been a response to a perceived problem (e.g., plane crashes due to pilot inexperience or the need for improved decision making). In the past few decades we have seen the advent of distributed simulation and the development of virtual environments as an alternative human-simulator interface.[3,4]

Figure 9-1   An early flight simulator circa 1910. A barrel was cut into two pieces and used to provide both roll and pitch. Note the "actuator" on the right providing the roll function.

---

[2] See *The Wars of the Jews or the History of the Destruction of Jerusalem*, Book III, Chap. 5, Sect. 1. (*circa* 70 A.D.) In this section Josephus describes the training of Roman soldiers in these words: "... their exercises unbloody battles, and their battles bloody exercises."

[3] See, for example, E.A. Alluisi. The development of technology for collective training: SIMNET, a case history. *Human Factors* 1991;*33*(3):343–362.

[4] See, for example, N. Durlach and A. Mavor (eds). *Virtual Reality: Scientific and Technological Challenges*. Washington, DC: National Academy Press, 1995.

CONVERGENT SIMULATIONS

A distinct thread in the recent evolution of simulation is **convergence**.[5] Historically, distinct simulation approaches (e.g., discrete and continuous) have been conceived and applications developed in a relatively independent manner. The demand for simulation applications that serve more than one audience and/or more than one purpose has led to the convergence of heretofore distinct simulation approaches. A modern example of significant interest to the military is the convergence of live, virtual, and constructive (LVC) simulation within the Joint National Training Capability (JNTC).[6]

Traditionally, the military has used live simulation as a primary means of training. With the high cost of such training and the growing shortage of adequate space for its conduct, constructive simulation and, more recently, virtual simulation have become increasingly important. The convergence of LVC simulation provides the military (and others) with the ability to "mix and match" simulation methodologies to meet both the training objectives of the commander and the constraints (time, space, cost) imposed by the training context. The ultimate goal of the JNTC effort is to deliver, for the commander, the needed training with the optimal mix of LVC any place, any time.

Efforts to develop convergent simulations are hampered by a number of issues. Among them is the culture of the simulation community. Lacking the cohesion common in many fields of science and engineering, "simulationists" come from many backgrounds and, more often than not, reinvent what others have done, exhibit the "not invented here" view of their work, and have not been widely exposed to the rich tapestry of simulation approaches. Other issues include the common ones of incompatible data formats and the lack of intercommunication between simulation components that have been developed independently without regard for the need to exchange data and control information. Yet another barrier is the human interface to the simulation (both in terms of operators and users). Many interfaces have been developed for specific simulation environments and then reused for other, perhaps related, simulations. Thus, the user community may associate a given user interface with a class of simulations and find it unacceptable when its context is changed or when another interface style is adopted.

SERIOUS GAMES

Modern computer-based games are often predicated on the marriage of simulation (and computer graphics) and entertainment.[7] Such games have been wildly

---

[5] This was a term used by the author for a presentation: Convergent Simulations: Integrating Deterministic and Interactive Systems. Human Performance and Simulation Workshop, Alexandria, VA, July 30, 1997.

[6] See http://www.jfcom.mil/about/fact_jntc.htm.

[7] Information on the Serious Games Summit, one of the current venues for discussing serious games, can be found at http://www.seriousgames.org/index2.html.

successful and have led to the production of very-low-cost delivery platforms that are characterized by high performance computing and graphics. Profits are derived from economy of scale, driven by the huge demand for products. Many within the military have watched this development and have recognized the potential of game platforms (and the underlying game engines) as a means of delivering, again at low cost, games that can provide some types of training. The challenges lie in 1) the insertion of appropriate instructional design methods into game development, 2) the demonstration of the efficacy of game-based training, and 3) recognition that some (many?) training applications, even in the military, may have limited audience size or require frequent updating.

It is interesting to note that the relationship between training simulators and entertainment is not really that new. The famous flight simulator (Figure 9-2) developed by Edwin Link in the early 20th century clearly connected his simulator concept to the entertainment systems commonly employed in carnivals.[8] In his patent Link stated, "My invention relates to a combination training device for student aviators and entertainment apparatus. ..." This device went to on make an immense contribution to the training of pilots during World War II.

A current example of a game-based training system is Pulse!! (Figure 9-3), under development by Texas A&M University-Corpus Christi and Breakaway Games, led by Claudia Johnston.[9] This game/trainer is using a commercial game engine to build scenarios that deal with trauma care. The graphics and capabilities of the system are on a par with the best games available today. Perhaps most important are the efforts being made to design the game with appropriate pedagogy and to evaluate how effective it is as a training system.

HUMAN-SIMULATOR INTERFACES

Flight simulation has typically used replicas of aircraft cockpits as the interface between the human and the simulation. Products like Microsoft Flight Simulator™ have, in the interest of accessibility, developed very high-fidelity simulators with interfaces based on the keyboard/mouse with the possible addition of other relatively low-cost interface devices.[10] Future simulations will most likely rely on **adaptive interfaces**—interfaces that are reasonably generic, at least within a given domain and that can be readily adapted to a range of simulations. To accomplish this capability, one needs access to a variety of displays: visual (three-dimensional, wide field-of-view), haptic, vestibular, olfactory, and (perhaps) gustatory.[11] In the intermediate term, we can expect display **devices** that couple directly or closely to the

---

[8] Edwin A. Link, Jr. United States Patent 1,825,462. Awarded September 29, 1931.

[9] See http://www.sp.tamucc.edu/pulse/home.asp.

[10] See http://www.microsoft.com/games/flightsimulator/.

[11] For a recent review of these technologies, see Dylan Schmorrow, Joseph Cohn, and Denise Nicholson, Editors. *PSI Handbook of Virtual Environments for Training and Education: Developments for the Military and Beyond.* Westport, CT: Praeger Security International, 2008.

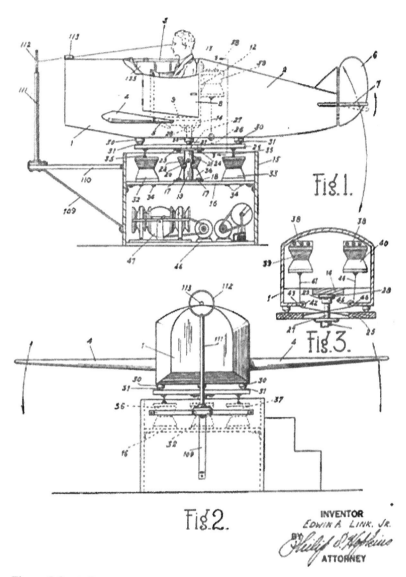

Figure 9-2   A diagram of the first commercially successful flight simulator, developed by Edwin A. Link, Jr. From US Patent 1,825,462

human sensory system. In the visual domain an example of this type of device is the retinal display, under development (and available in limited capability versions) from Microvision.[12,13] The concept is simple: use one or more color lasers and a raster device to "write" images directly on the human retina. Sophisticated, but

---

[12] See, for example, http://www.cs.nps.navy.mil/people/faculty/capps/4473/projects/fiambolis/vrd/vrd_full.html.

[13] See http://www.microvision.com/.

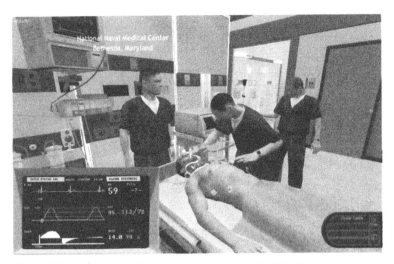

Figure 9-3   National Naval Medical Center as modeled in Pulse!!

highly constrained, haptic displays are also commercially available.[14] Vestibular displays (e.g., motion bases) are also available as are a few examples of olfactory displays.[15,16] While serious engineering is still needed, there is the potential for these displays to mature to the point where their cost, ease of integration, and robustness is sufficient to serve as interfaces to simulations in a variety of application areas.[17,18]

In the longer term, we will see the introduction of the means to directly connect to the human sensory system. At least one example has been in place for some time—a vestibular display developed by the U.S. Air Force.[19] This display directly stimulates the human vestibular system. One can conceive of a family of such

[14] See, for example, http://www.sensable.com/.

[15] See, for example, http://www.inmotionsimulation.com/.

[16] For pointers to examples of olfactory displays, see Donald A. Washburn and Lauriann M. Jones. Could olfactory displays improve data visualization? *Computers in Science and Engineering* 2004;6(6):80–83.

[17] For a somewhat more discursive discussion of multi-sensory display technologies, see R.B. Loftin. Multisensory perception: Beyond the visual in visualization. *Computers in Science and Engineering* 2003;5(4):56–58.

[18] See also, C Basdogan and RB Loftin. Multi-modal display systems: Haptic, olfactory, gustatory, vestibular. In *PSI Handbook of Virtual Environments for Training and Education: Developments for the Military and Beyond*. Edited by Dylan Schmorrow, Joseph Cohn, and Denise Nicholson. Westport, CT: Praeger Security International, 2008.

[19] JD Cress, LJ Hettinger, JA Cunningham, GE Riccio, MW Haas, and GR McMillan. Integrating vestibular displays for VE and airborne applications. *IEEE Computer Graphics and Applications* 1997;17(6):46–52.

devices that are directly coupled to the human sensory system and that, in principle, could provide the ultimate display capability (with all due respect to Ivan Sutherland).[20]

COMPUTING TECHNOLOGY

Over the past 20 years we have seen an extraordinary evolution of computing capability. Nonetheless, the simulation community has often been ahead of the curve. That is, as soon as we can simulate, in real time, 100k entities, the community demands that we simulate 500k or 1000k entities with the same speed. The current frontiers of computing technology are focused on quantum computing and biological computing.[21,22] In either case, there is the potential for several orders of magnitude improvement in computing capability with a concomitant reduction in the size and power requirements of the devices. This opens the opportunity for essentially unlimited entity count and performance on the compute side while enabling simulation to become truly portable.

THE ROLE OF EDUCATION IN SIMULATION

Recognized disciplines (e.g., Physics, Electrical Engineering, Computer Science, etc.) have strong mechanisms that create identities for "practitioners." Those who call themselves Electrical Engineers, for example, have shared a fairly common undergraduate experience. They take many of the same courses, regardless of the university attended, touch many of the same textbooks, receive a degree from an academic department within an institution, and will, perhaps, take a licensing examination which varies little from state-to-state in the United States of America. In doing these things they are exposed to a common body of knowledge. In some cases (Computer Science) an explicit description of this body of knowledge exists.[23] In other cases it is implicit in key textbooks that are widely used by undergraduate students of the discipline. Basically, during their undergraduate experience, students are acculturated into a discipline. They begin to speak with the same technical terms, hear many of the same stories about key events in the history of their discipline, and lionize the individuals who have made major contributions to the field.

---

[20] Sutherland IE. The Ultimate Display. In Proceedings of the International Federation of Information Processing Congress 2, pp. 506–508 (1965).

[21] See, for example, Mika Hirvensalo. *Quantum Computing*, 2nd Ed. Berlin: Springer-Verlag, 2004.

[22] See, for example, Leandro N. de Castro and Fernando J. Von Zuben. *Recent Developments in Biologically Inspired Computing*. Hershey, PA: Idea Group Publishing, 2005.

[23] http://www.sigcse.org/cc2001/cs-csbok.html.

Simulation lacks almost all of these characteristics. There are no "Departments of Simulation" at universities. There are no undergraduate degrees in simulation. There is no (until now) comprehensive undergraduate textbook that is widely used. While there are stories and "lions," they are not all well known among those who practice simulation. In short, there is no simulation "body of knowledge." Instead, individuals come to simulation from a variety of backgrounds. They may be mathematicians, physicists, chemists, computer scientists, or engineers (of all stripes). In almost every case, the person practicing simulation will likely refer to themselves as a member of the discipline from which their undergraduate degree comes. At this writing a few universities (University of Arizona/Arizona State University, the Naval Postgraduate School, Old Dominion University, and the University of Central Florida, for example) offer masters and/or doctoral degrees in Modeling and Simulation. The "fodder" for their programs, however, comes from established disciplines and the nature of their graduate programs does not serve to acculturate them into this new discipline nearly as well as their undergraduate major did to the discipline inked on their diplomas. We do have a professional certification, but we lack government and industry mandates that bring real value to the certification (as is true in engineering).

What do we need? The field needs a book like this one, and it needs an undergraduate degree. In creating these two attributes, a consensus will eventually emerge on a body of knowledge. Ultimately, these factors will give birth to another discipline that will stand alongside those mentioned above, creating a culture of simulation and a generation of graduates who think of themselves, first and foremost, as Modeling and Simulation professionals.

THE FUTURE OF SIMULATION

How can we envision the influence of developments, along the four dimensions explored above, in the creation of a future simulation? First, let us focus on a specific application of simulation: training. With this in mind, we must think in terms of a seamless mix (or convergence) of live, virtual, and constructive elements with the fewest participants in the live and the most in the constructive domains of the simulation. However, *from the users' perspectives* they will not be able to identify which entities are real participants in a real-world setting, which are real participants in a virtual environment, and which are computer-generated entities in a virtual environment. Second, we can assume that we will take advantage of advanced computing technologies and serious gaming techniques that provide the real participants in the virtual environment with a world "virtually" indistinguishable from the real world in terms of its sensory fidelity and interactive responsiveness. Third, the users will interact with both virtual environments using interfaces that are tightly integrated with their own sensory systems and that do not intrude in ways that render the virtual world any less believable than the real world. Finally (and again) computing technology will allow essentially unlimited entities in both the virtual and constructive components while enabling truly adaptive capabilities in terms of the evolution of the simulation in response to user actions. The resulting simulation will, if properly designed and executed, provide the ultimate in training efficacy, any place, any time.[24]

KEY TERMS

convergence
adaptive interfaces
devices

[24] The author was tasked, over 16 years ago, by Mark Gersh, then at Headquarters, National Aeronautics and Space Administration (NASA), with developing a white paper on the "future of training." I am indebted to Mark for the initial stimulus to think about such things as directly coupling humans to simulators. I am also grateful for the support of Robert T. (Bob) Savely of the NASA/Johnson Space Center in carrying out that task. Most of all I must acknowledge Old Dominion University for giving me the opportunity to shape its graduate programs in Modeling and Simulation and to work with outstanding students who challenged me to help them become Modeling and Simulation professionals.

Index

Principles of Modeling and Simulation: A Multidisciplinary Approach, Edited by John A. Sokolowski and Catherine M. Banks.
Copyright © 2009 John Wiley & Sons, Inc.

Printed and bound by CPI Group (UK) Ltd, Croydon, CR0 4YY

16/04/2025

14658517-0001